To Daniel

your father really

loves you...!!

Good Luck!

Betty

Praise for
The Great Digital Transformation

"I've known Gerry Szatvanyi for more than ten years in various roles centered on digital transformation. He has walked the road to transformation and has stories, advice, and best practices to share, best practices gathered through working with many world-class companies. In his book, you will find lots of examples and wise pieces of advice to understand more about what is essential to succeed in a changing world"

—Guy Gauvin, Chief Operating Officer, Coveo

"Brands and retailers must digitally transform their customer interfaces to survive and thrive. These same consumers experience innovation driven by cutting-edge apps and services in media, entertainment, communication, education, and more. They expect and require the same level of customization, ease of use, and pleasure in their commerce experiences. Their patience for friction is very low. Gerry Szatvanyi and OSF Digital have been at the forefront of helping many global brands innovate their methods for digital interaction with customers. Digital transformation is Gerry's strength. His technical expertise coupled with a worldwide customer base innovating across customer experience has informed his expertise and knowledge. This book does a great job of sharing those insights in an enjoyable read."

—Rand Lewis, Cofounder and Managing Partner, Delta-v Capital

"*The Great Digital Transformation* is a must-read for any executive who's trying to understand the practical realities of a digital transformation. Gerry Szatvanyi, CEO of OSF Digital, has been at the forefront of the evolving digital revolution for companies, customers, suppliers, and employees. His lessons learned from working with the world's biggest brands are essential reading for anyone driving meaningful change to the way brands are experienced in the modern world."

—Jeff Rich, Managing Director, PlumTree Partners

"The startups that I've led in the past were all about digital transformation, specifically with AI and IoT. My collaboration with Gerry allowed me to have an insight into how Gerry thinks, and I can say that his book is the culmination of his personal experiences in hands-on operation, disruptive innovation, and strategy. Gerry is an amazing advisor on the new technologies you must understand to give you the confidence to ask the right questions and drive change that delivers both short-term results and long-term competitive advantage."

—Cyril Brignone, Cofounder, Vecitice

"eSkill is all about digitally transforming the recruitment selection process, going from paper or less-performant systems to state-of-the-art assessment tech. As CEO of eSkill, I've collaborated with Gerry in leading this digital transformation, which has enabled our clients to accurately assess millions of candidates globally and digitally transform their talent management processes."

—Eric Justin Friedman, Founder and CEO, eSkill Corporation

"If you are planning or already executing a digital transformation then you have to read *The Great Digital Transformation*. Authored by Gerry Szatvanyi, the book is a reflection of his real-world experience leading companies through the digital transformation journey. Gerry is a technologist, entrepreneur, and leader who was driving digital transformation even before the term itself was commonly used. So take advantage of Gerry's years of experience to enhance the journey for yourself and your company."

—David Northington, Board Member, Adviser for High-Growth Tech Companies

"This book is a must-read for anyone engaged in a digital transformation vision and strategy. Over the years, Gerard and OSF have worked with world-renowned brands and helped them reinvent their customer engagement model to meet customers exactly where they are. This book is packed with actionable insights from digital and commerce gurus."

—Fabrice Talbot, VP of Products, Salesforce

"I had the incredible chance to meet Gerry Szatvanyi more than a decade ago. Starting to work with him and his OSF team on our first flagship ecommerce sites, Gerry quickly became a true business partner and a friend. Having a sincere consumer-centric approach and constantly challenging my team and I to reinvent ourselves, he was a cornerstone for our enterprise digital transformation program. Gerry is a passionate and visionary digital entrepreneur full of common sense. No wonder why his enterprise OSF developed at such a pace. His book is full of practical ideas and is definitively a must-read for anyone who wants to engage his business in a true digital transformation."

—Rejean Demers, CIO Global Travel Retail, L'Oréal

THE GREAT
DIGITAL
TRANSFORMATION

THE GREAT
DIGITAL
TRANSFORMATION

1 0 0 1 0 1 1 1 0 1 1 0 1 1 1 0 0 1 0 1 0 1 1 1 0 0 1 1 0 1 0 1 1

Reimagining the Future of
Customer Interactions

GERARD
SZATVANYI

Forbes | Books

Published by Forbes Books, Charleston, South Carolina.
Member of Advantage Media.

Forbes Books is a registered trademark, and the Forbes Books colophon is a trademark of Forbes Media, LLC.

Printed in the United States of America.

10 9 8 7 6 5 4 3 2 1

ISBN: 978-1-95588-445-7 (Hardcover)
ISBN: 979-8-88750-109-3 (Paperback)
ISBN: 978-1-95588-483-9 (eBook)

LCCN: 2022917374

This custom publication is intended to provide accurate information and the opinions of the author in regard to the subject matter covered. It is sold with the understanding that the publisher, Forbes Books, is not engaged in rendering legal, financial, or professional services of any kind. If legal advice or other expert assistance is required, the reader is advised to seek the services of a competent professional.

Since 1917, Forbes has remained steadfast in its mission to serve as the defining voice of entrepreneurial capitalism. Forbes Books, launched in 2016 through a partnership with Advantage Media, furthers that aim by helping business and thought leaders bring their stories, passion, and knowledge to the forefront in custom books. Opinions expressed by Forbes Books authors are their own. To be considered for publication, please visit **books.Forbes.com**.

TO MY TEAM.

#OSFSTRONG

CONTENTS

ACKNOWLEDGMENTS

Thank you, Momo, for everything you did for me. Mityu, without your guidance, my journey would have not started. Adela, thank you for always supporting me. And Luana, for being my wife, my partner, my friend, and my confidante, thank you!

INTRODUCTION

With today's technology, the world is at our fingertips.

Do you agree with that opening statement? Do we really have access to just about anything, provided we have a device in our hands?

Perhaps to a certain extent, especially thanks to innovations in the last decade and advances that help connect businesses and consumers, we might state that many goods and services are universally and abundantly available today. If we dig deeper, however, and pan out the lens to view the wider context of areas like commerce, I think we'll find a different answer. Allow me to provide an illustration to explain.

Let's follow a traveler through a Chicago airport. She's flying out of O'Hare after spending a weekend touring the Windy City. She returns a rental car she used during her visit at the car rental lot just outside of the airport. Then she hops on a shuttle that will take her to her terminal. During these first legs of her trip, she doesn't hear or see anything on the shuttle bus that alerts her to all the stores and offers within the airport.

Once checked in and through security, she heads to her gate. She sees a convenience store and stops to pick up a few snacks. While the cashier behind the counter is ringing up her purchases, the traveler spots a display of T-shirts with the word *Chicago* and the skyline painted on them. In that moment, she realizes she hasn't purchased souvenir attire yet and wants a shirt as a token memory of the city. Unfortunately, as she browses the options, she doesn't see her size. When she asks the cashier for help to find a size medium T-shirt, the only reply she receives is, "If you don't see them here, we don't have any. I don't know if any of our other stores carry them."

The traveler leaves the store and heads to her flight, her snacks in hand. Notice that she carries only food, despite the fact that airports are turning into veritable shopping malls.

Did you catch the missed opportunity of a shirt sale? In that moment, is the world really at our Chicago tourist's fingertips? And perhaps more importantly, what other losses can be found in this tale?

I would argue that the misses here go well beyond a disappointed customer. We don't know what type of shopper she is, her past purchases at the store, what she might do if the item could have been shipped to her, what she would buy if the store had an online presence that was connected to the in-store experience ... and on and on.

And yet I'm here to assure you that it doesn't have to be this way. In fact, technologies are readily available and accessible today that solve the problems and barriers that can easily separate customers from stores and result in failed transactions. The world is connected, and the right solutions can be put in place to better understand customers, improve their shopping journeys, and create a higher number of transactions.

At OSF Digital, we've had an inside seat, to a certain degree, to watch these changes in commerce unfold. As the world has grown more connected, we've adjusted too. When we first started in 2003, we built digital products for clients to help them better engage customers: think implementing websites and transferring pen-and-paper-driven tasks to the digital space.

Today, just as putting up a website alone won't automatically save a sinking brick-and-mortar store, the work we carry out at OSF Digital is much different from what we did ten years ago. We've seen that merely putting previously nondigital processes into a digitized state isn't enough. Retailers and brands today—commerce in general, if you will—need to adapt to a new shopping world that has many facets, including different types of buyers and shopping patterns, along with ever-changing ways to experience shopping.

It means rethinking how we do business and changing models to be more appropriate for certain technologies and customer expectations. Along those same lines, businesses that want to perform well in the future—even if they are doing well now—are ripe for taking some time to spot these new opportunities. It's a chance to evaluate current methods and to evolve into a new way of being.

It's not that we've thrown out our old ways. At OSF, we still very much focus on digital products and we're still building digital platforms for clients around the world. We've been a trusted Salesforce partner for more than ten years and are certified on all clouds. We've helped both B2B and B2C companies and have provided technology, consulting, implementation, and online shop management services to both up-and-coming and premier brands as well as to merchants focused on building multicloud and unified commerce projects using Salesforce and other top-tier cloud technologies.

In our work today, though, we now focus much of our time on helping clients with digital transformation.

Let's pause for a moment and reflect on digital transformation. The term tends to be a buzzword, but its definition and its true impact aren't always as readily clear. That's why in the pages that follow, we'll drill down and consider exactly what digital transformation consists of. We'll see how companies are truly revamping the way they interact with their customers, how their employees work, what purpose their stores serve, and more. In doing so, we'll explore new possibilities that can help organizations stay connected to customers and thrive.

In recent times, we've seen significant changes in the world around us, and rest assured, we're not going back to how things were. But that doesn't have to be a bad thing or cause concern. Quite the opposite, in fact. This new world we live in, full of remote workers, mobile-minded customers, and many empty stores presents nearly endless opportunities for those who are willing to dream. Right before our eyes lies the opportunity to deliver services in a new way, to break into new markets, and to increase market share.

I invite you to sit back, close your eyes, and mentally wipe away what is considered to be standard today. As you read through the chapters in this book, allow yourself to be creative. Ask questions.

Another example can help demonstrate the potential before us. Let's toss out the currently accepted healthcare system that exists in many locations throughout the United States. The traditional model consists of doctors working in offices from nine-to-five. Each day, patients who live in their area travel a few miles by car, bus, or foot to see the doctor in person and receive treatment and health services.

Why does that office exist? Why do those doctors and nurses work the hours they do? Taking it a step further, why are these professionals only serving patients in their community?

As we've seen, especially with the onset of the pandemic, we have the means to virtually meet with anyone from anywhere in the world, provided we each have the necessary equipment and connection to do so. Taking this point and plugging it into the healthcare system can create a whole new vision of how doctors and patients interact. I challenge you to answer this question: Why would someone living in Montana, a highly rural western state, not be able to receive the same level of service as someone living in Manhattan, one of New York City's finest and most upscale boroughs? If we truly can use technology in the ways we know exist, we can open gateways to build all-encompassing, better-reaching healthcare systems.

Or consider the watches we wear, and connect that commodity to the concept of healthcare and the power of prevention. If you wear a watch today, it's likely it can do many things beyond telling time. It might monitor your heart rate and calculate how many steps you walk during a twenty-four-hour time span, for instance. So why is that same watch not monitoring your blood sugar levels and other crucial health components, like digestive functions? That system could be connected to someone on the other side of the world, where it is monitored, and if something out of sync is noticed, a person could call you and say, "Okay, let's sit down and have a conversation. Let's talk about what's happening here."

Even if you're nodding your head or agreeing with this idea, don't expect to wake up tomorrow and see a brand-new healthcare system. You probably are not planning to Zoom in with a doctor at the Mayo Clinic, one of the world's top medical centers, from your residence in South Africa for a routine checkup (unless you have already established a connection through an in-person visit). And there may not be anyone monitoring your heart rate and other vital signs through your watch (at least, not that you know of!). No, today's healthcare system is strongly rooted in the way it currently operates. It's much easier to go along with what's always been done before or to look for ways to improve efficiencies within the systems we already have in place.

And yet we need to account for timing. There is a sense of urgency surrounding these changes, and rightly so. Technological advancements during the 1980s, the 1990s, and the early millennia provided a turning point for growth. The advent of the internet brought great opportunities and disruptions, and the adoption of mobile use has created a new way of living and interacting. Moreover, the combining of technologies and their ongoing evolvement has increased the pace of change. For companies that resist adjusting (and it's so easy to do), there is great risk of becoming obsolete in the next years—if not months. When competitors start offering these new ways of delivering services and branching into new marketplaces, those who are left scrambling to develop similar offerings will be starting the race from behind. Moreover, companies that are performing well need to pivot and adapt now to ensure leadership in the years to come. They'll want to set the pace for others to follow rather than encounter obstacles that could deem them irrelevant.

But that's not all. The fast track of change is not going to slow down. Transformation is still happening in a big way, and it will continue to do so in the coming years. How can you tell we're still in the middle of the transformation? Just walk into a store, and you're likely to see that all the people in the shop are either at the cash register or putting inventory on the shelves. If they are not interacting with the customer, something is not right.

They haven't thought through what the problem is. Did you notice it? The issue is that they have humans performing repetitive tasks that machines can carry out. You don't need a line of cashiers when you have self-checkout. Why, then, is the customer like our

Chicago traveler left alone to guide themselves through the store? We need to shift from a transactional mindset to one that is focused on interaction with the customer. In the case of our Chicago tourist, it could mean that she leaves the airport with more than just snacks; she could have a new T-shirt in a shopping bag, thanks to the shopping clerk's ability to check the inventory of other store locations within the airport to find the size medium souvenir. The traveler might even have learned, while on the shuttle bus after returning her rental car, that two of her favorite brands have small pop-up locations inside the airport and that she can reach them if she makes a very slight detour while on her way to the gate. If she looks online at those stores and sees a few items she wants, she can ask about them at the pop-ups. Perhaps one of the locations has her requests; the other doesn't but agrees to ship the merchandise to her home address at no cost and notifies her that it will arrive within five days. At every stop in the airport, the places gather a few details about the traveler, which

> **We need to shift from a transactional mindset to one that is focused on interaction with the customer.**

she readily shares because she feels confident that they'll send her promotions that fit her current interests and style. They also offer to send her a few additional items from time to time, giving her the choice of trying them out and deciding whether she'd like to purchase them or send them back.

To make these changes, we don't need earth-shattering new technology. We don't need to build systems from the ground up and code our way to a massively disruptive system. There's no R&D department required to study this new trend. Quite the opposite. These technologies are there and ready to be implemented. We just need to know how to best use them for our purposes. It starts by sitting down, clearing our minds, and thinking of different ways to help customers find what they need. What you need are dreamers—people who can reimagine your experiences.

So let's dream. And then let's make it happen.

CHAPTER 1

The Difference between Digitization and Digital Transformation

True transformation begins with a step back and an invitation to rethink how we live and work.

I recently walked into a retail store to get a new pair of shoes. Not only did the shop not have my size but also there was no one around who was actively engaging with customers. I didn't find the item I was looking for, and I received no assistance to point me to where I might find it, how a pair could be ordered for me and sent to my home, or what else I might want to buy while I was in the shop. I left empty-handed and without talking to a single clerk … no dollars spent.

Was it just bad luck?

Perhaps, but I ask you to think back and consider when and where you may have had a similar experience. The reality is that many retailers are operating in this way today. They may or may not have the inventory their shoppers are seeking. They might have employees standing behind the counter at the checkout in the front of the store, waiting to make the transaction.

But today's customer doesn't necessarily need help carrying out a transaction. The purchase can be made with just a few clicks online. Or by checking out at a self-service counter.

When I see scenarios such as this one, it makes me think about what could be done to improve the experience. For instance, what if employees were available to guide shoppers through the store and act as an advisory for shoes and other merchandise? Those same workers could have access to online systems that could check inventory levels. An arrangement might be set up in which the store offers at least one pair of shoes in every size. That way, consumers (myself included) could be led to the shoe section, where we could try on the footwear and then have the right size sent to our home if it's not in stock at the store.

Taking the example one step further, what if workers who were not busy helping clients in a store had access to online systems and could provide assistance to someone shopping online who asks for help? The employee could log in and talk to customers in an interactive, engaged way. The very working environment could be arranged so that employees would carry out store tasks as needed; during their downtime they would provide online support for shoppers.

Consider a scenario in which a customer could access a store assistant through a video call if needed. Perhaps all the store representatives are helping other shoppers, and a customer is lingering over a question about a purchase they may or may not make. That customer could use a video connection in the store to access an assistant in another location. This arrangement could be similar to other areas in our lives in which we use remote technologies to carry out consulting engagements. Think of how a personal trainer doesn't have to be present with you every time you work out; instead, the two of you can connect via a video meeting to go over your latest regimen. Applying this to retail, shopping assistants in store locations on the West Coast could help shoppers on the East Coast during high-traffic times in the region, and vice versa. This could all be done via video technologies, which give the customer the chance to see and hear the representative and view any available merchandise they are discussing.

By imagining a situation like this, it not only shows us ways to be more productive with employee time and develop an improved customer experience but also presents the chance to elevate employees themselves. Workers might not be as interested in standing at the counter and scanning items repeatedly for eight hours straight. They may very much enjoy being more involved and having greater interaction with the customer. They might feel empowered when they act as a business partner, leading the customer through the experience and upselling items or pointing out other goods the customer might not have considered but would find beneficial.

This example of shopping for shoes—and then envisioning new ways the process could unfold—demonstrates several trends that are underway in our world today. Let's spend some time looking at these shifts in commerce, along with what it means to digitize and go through a digital transformation. The exercise will enable us to think beyond traditional formats and consider new environments that are more inviting for customers, workers, and companies.

The Downfall of Retailers

Have you ever driven or strolled past shops you once frequented but that now have a Permanently Closed sign on their door? Or walked up to a place, only to realize that another company had moved in? It's not uncommon to see stores that we grew to love in earlier years close their doors or cease to exist.

Perhaps a glance at recent headlines sheds more light on this troubling tale. After the onset of the pandemic, we watched as brands such as J. Crew, Neiman Marcus, and JCPenney filed for bankruptcy.[1] COVID-19 certainly intensified ongoing issues such as financial constraints, and it caused disruptions that, coupled with mandates to close stores for health reasons, unraveled a frightening number of companies. The pace of retail failures broke records during 2021.[2]

Problems plaguing retailers actually date to prepandemic times and coincide with changes in technology and the way we live and work. Sometimes referred to as the retail apocalypse, the number of brick-and-mortar retail stores, especially those of large chains that operate globally, began declining around 2010.[3] In 2017, a Business Insider report called the phenomenon the Amazon effect and estimated that Amazon.com was generating more than 50 percent of the growth of retail sales.[4] In 2019, retailers in the US announced more than 9,000 store closings, which signaled a 59 percent jump from 2018.[5] It's unsettling to think of what awaits the brands that have—thus far—survived.

We could make a long list of the reasons why these retailers didn't make it. When I survey the evidence, I come to the conclusion that, in many cases, the downslide started when the way we live and work shifted. Rather than moving into this new era, many companies continued to carry on and do things the way they had always been done. While continuing down the same path, they expected different results. Unfortunately, those different results never came.

Before we think too harshly about retailers that have gone bankrupt, let's acknowledge that it's a changing world, and it's tough to keep up with the competition. Some, if not all, of the failed brands we've seen in recent years made an effort to innovate, albeit at times it might have been too little, too late. The underlying cause of their failed attempts typically stemmed from trying to improve processes but not change the way the systems were set up. In some cases, they were merely digitizing old ways.

What Is Digitization?

Digitization refers to putting information in digital form. If we are implementing technologies—for instance, shifting away from paper processes and making them digital—we might say we are undergoing digitization. There can be an aura of advancement around these upgrades, and indeed, they may be time saving to a certain extent. The issue arises, however, when these processes follow the same path or keep departments in silos. They don't make it possible to better integrate and streamline processes. In this way, they have certain limits to them. You could digitize a method of doing a task, but the new process still serves the same purpose as before and doesn't create a substantial change to the business approach.

By itself, digitization can provide great benefits. But it won't necessarily take us to a truly different model that puts the customer first and breaks down silos. The reason for this is most frequently because retail today is very different from what it was ten years ago. The way consumers interact with brands is evolving, and e-commerce and commerce are merging. Not only that, but this trend is one that we can expect to continue for the coming years. In a sense, this wave of new commerce has started but not yet crested. And even after it crests, it will roll into a new form and shape that will look much different from what we now see. To get to a new experience for all involved, we need to go through a digital transformation.

What Is Digital Transformation?

Digital transformation may be tossed around in today's headlines, but it has substance behind it. Digital transformation, first and foremost, is about dreaming … about taking new technologies, new methods, and recombining them in different ways to reimagine your business and the way your business really interacts with your customers. In commerce, digital transformation holds the possibilities of a new shopping experience, better control over data, and a chance to reimagine how products and services are sold. It also involves revisioning the working environment for employees and associates, a deep understanding of the customer journey, and a better picture of what is going on in stores, online, and everywhere else.

It's important to recognize that times are changing. If, a decade ago, we were implementing new technologies such as selling online or setting up a website, that might be called a transformation. The issue, then, is that these technologies and with them, the capabilities, have evolved in the last years. This makes the concept and the expectation change. In a sense, these advancements that were once considered to be cutting edge have become basic practices in today's world; hence, you can no longer think of them as a true transformation.

IT'S NOT ENOUGH TO SELL JACKETS ONLINE

Let's say a company produces and sells winter coats. The firm has been in existence for fifty years and has done well; it can boast fifteen different locations, along with a headquarters. Its retail stores have an avid customer base, and sales have steadily moved upward. Suppose you heard about the company on a radio station commercial. The ad said you had to go to the location to make a purchase. The store did not operate online at all.

Perhaps the board members of the chain of stores hold a meeting, and they decide to put up a website. The new site will allow customers to browse the merchandise and make a purchase that will be delivered to their home. Shoppers will even be able to start the return process if they want to send back an item.

In this case, even after the website is up and running, the company is not transforming. It is simply putting processes that it already has established into digital form. We could say it is going through a digitization but not a digital transformation.

The definition of digital transformation itself has significantly evolved over the years. Digital transformation calls for a step back to imagine the possibilities you have before you today. Keep in mind that commerce now takes place via mobile technologies, which include social networks, video calling, and even online gaming formats. Many of these have only been developed and made available to consumers in the last few years. Consider other breakthrough technologies such as augmented reality, which greatly impacts the way we interact. Augmented reality consists of putting a computer-generated image into the world

> Digital transformation calls for a step back to imagine the possibilities you have before you today.

that the user is viewing. Think of looking at an actual beach scene. You are standing in the sand and holding your device so that the water is displayed on the screen. You then virtually place a boat you are thinking about buying on the water. The boat appears in the image you're viewing through your device.

USE CASE: BURTON SNOWBOARDS DIGITALLY TRANSFORMS FOR TODAY'S CUSTOMER

Burton Snowboards manufactures and sells snowboarding gear, apparel, and related products. The company's product list extensively covers the snowboarder's needs. The company, based in Burlington, Vermont, offers snowboards, boots, bindings, tools, and accessories, along with backpacks, shoulder bags, travel bags, luggage, snowboarding gear bags, and specialty bags.

Burton markets their products through their own stores. The firm also sells through their partners' stores and other online shops in both North America and Europe.

Founded in 1977, during its early years, Burton sold its products almost exclusively through wholesale channels. As time went on, about 75 percent remained wholesale, and 25 percent of its business came in the form of direct-to-customer (DTC) sales. While the DTC side of the business had the potential to grow significantly, the company didn't have the systems it needed to support the segment.

With the goal of growing their DTC business, Burton launched into a digital transformation to become a consumer-centric, digital-first sporting goods experience company of the future. The company wanted to revamp their e-commerce and change the way they interacted with customers. Rather than providing a purely transactional engagement, in which the customer picked out a snowboard or other merchandise, paid, and disappeared, Burton wanted to develop a deeper connection. The company aimed to provide an experience: a touring of sorts for the customer of what was available, the different ways they could personalize a snowboard to make it fit their personality, and, perhaps most important, a reason to return.

To undergo this transformation, Burton focused on developing a community. The company recognized that snowboarders in general tend to be aware of how they look when they are outdoors—and they are very interested in how their snowboarding equipment fits into that image. Customers were looking for more than a payment and delivery; they wanted a design that spoke to their personality and a place where they felt they belonged.

Now when a new customer lands on Burton's site, they can compare snowboards and snowboarding equipment and attire. They can choose the colors they like, add a design, and personalize the entire purchase to suit their preferences. Burton tracks all of this so that even after the sale, the company can stay in touch with the customer and cater future promotions to their past purchases. The company also keeps careful tabs on customer data to see how product lines are performing and what else might be interesting to their customer base.

The changes that Burton implemented put customers first. The adjustments also had additional benefits. Thanks to these processes and the sense of community the company created, it became easier to do more cross-selling and upselling. For instance, when a customer selects a snowboard, they can choose if they'd like to put their name on it. As they indicate their design and style preferences, the customer can be offered gear that fits their taste, such as boots, gloves, or a jacket to go with the snowboard. And after the sale, there is an opportunity to reach out to service the board, offer a chance to see new products, and bring the customer back to have another experience—and then join other Burton owners on the slopes.

Changing How We Work

The idea of real transformation is usually associated with the concept of disruption. It doesn't overlook the fact that many workers are forming a new preference for the way they earn income. Employees today (and even more so tomorrow) want to punch in and punch out differently. For instance, think of Uber employees. Once they own a vehicle and license and have passed the approval process, they have a flexibility that would not have been possible at a taxi company fifteen years ago. There is no nine-to-five with Uber. Instead, if you want to work two hours one day and then five hours the next, you're able to do so without anyone asking you why you didn't put in more hours. You have the freedom to choose when you want to work.

Some Uber drivers might use their driving time as their main source of income. Others, however, will use the work as a side job to bring in some extra dollars. They might have two or even three part-time jobs. They put these together to create the type of lifestyle that they want. In this gig economy, we're moving away from the eight-hour on-site shift and drifting to a system that is more fragmented and that yet allows for more possibilities. What if you had an employee who worked for two hours but who carried out the same amount of productivity as one who worked eight hours?

Could that work style translate to other industries? I invite you to think about your own organization and area of expertise. Consider that workers are looking for these benefits when they take on a new job. Take some time to think about how workers could be given more flexibility and a greater sense of purpose.

Meeting Different Customer Preferences

Digital transformation enables us to recognize different shopping patterns. Some shoppers, for instance, might want a subscription service so they can always be assured they'll easily get the same item they want again and again. Think of a man who buys a polo he loves. He may want that exact shirt to show up at his doorstep several times a year or to be able to buy it in five different colors. A subscription service could send out the same items at set times, or it could send out new options to consider. The items that the customer doesn't want can be sent back, free of charge. The whole process could start with a card inserted in the packaging that notes, "Would you like us to ship you the same shirt again? Simply follow these instructions ..."

Then there's another type of shopper who will be interested in browsing and considering recommendations, reading reviews, and constantly trying new things. Think of the working mother who just moved to a new home. She wants to decorate it, furnish it,

> **Take some time to think about how workers could be given more flexibility and a greater sense of purpose.**

and get settled. After that, she will continue to improve and enhance its look. In general, she is open to more suggestions and ideas. That demographic of shopper will fall into a category of different expectations and needs and will do well (i.e., spend more and be more satisfied) if the system can provide what she is looking for.

On a more upscale level, there may be someone who purchases from a retailer or brand a couple of times a year. A sales associate could look at the data of their buying patterns and occasionally reach out to the consumer, inviting them to the store or asking if they'd like to consider some new inventory that just arrived. A Zoom session or videoconference could be set up to see what the customer is interested in and how the store can help provide that. Selling more to the customer might not require a steep discount; on the contrary, if the customer has already made a purchase and is satisfied with having items curated for them, it's possible that no discount will even need to be offered.

Lost in the Transaction

Consider the customer who makes a large transaction and then is never contacted by the store again. Why is there a lack of communication? Why is their data not available to be seen and followed up on? An employee could look at their transaction and then reach out to invite them into the store. The sales associate could prepare items for the customer to consider and then spend thirty minutes with the shopper to go over some options. In this case, when the customer arrives at the store, items in their sizes are already set aside, and they don't have to go through the departments searching for things that have the potential to be a good fit. Instead, they can arrive, be guided by an advisor, try on items, and leave with whatever they want. As they move through the store, there will be opportunities to cross-sell too.

Tracking customer purchases has another benefit: the supply chain. You can pay attention to what is being bought, how often it is purchased, and what goes into the sale. The process allows you to avoid manufacturing items that have a lower chance of selling. Consider this: Rather than taking the risk of too much inventory, you have the opportunity to make the most of your manufacturing facilities and suppliers. It creates the chance to become more in tune with what's happening in the market, with what people want to buy, and the experience they want to have.

These are the opportunities that digital transformation brings.

PUTTING IT INTO ACTION

To start your own journey of digital transformation, ask yourself these questions:

- Why do we have the processes in place that we do?
- What could we do differently if the sky were the limit?
- Is new technology unlocking new ways of doing things? Think 5G and how it enables video calls, chats, and remote assistance.
- How can we take that sky-is-the-limit vision and apply it to our parameters?
- How could we put the customer first and be experience minded?
- What are our goals for the short- and long-term future?
- How can we implement changes to get there?

CHAPTER 2

Digital Transformation and Talent

You cannot do digital transformation and not rethink how
you work. You have to reimagine the company's culture
to ensure that innovation is supported, not stifled.

Let's take a moment to imagine a scene together; it begins with us
walking into a store. Say it is a technology retailer, a national brand
with locations throughout the country. This brick-and-mortar place
is in a prominent location and has been in this same spot for the past
ten years. Customers know where to find it and overall have an idea
of what it offers. We walk through the sliding doors and are greeted
with bright lights, white walls, and a clean environment.

It's a weekday afternoon, and the store isn't overcrowded. We stroll the aisles, which—for the most part—seem well stocked. I'm looking for a particular gadget that I saw in an ad online. It is a microphone that is supposed to work well in any location, regardless of background noise and movement. I saw that the retailer listed the microphone on its website. I wanted to see it in person before making a final decision.

Now in the store, I don't see the microphone, even after browsing areas with similar merchandise. We spot a store associate unloading boxes and filling shelves. We get his attention and ask him for help. The employee listens to my request and then heads to a workstation to check the digital inventory system. He discovers the item is in the back. He heads off to retrieve it and deliver it to me.

When the store associate places the package holding the microphone in my hand, it looks exactly like what I saw in an ad. However, I have a few questions about how to use it and how much it costs. I'm also curious to know if there are headphones to go with it. I ask for his recommendation for headphones and if a warranty is available (or worth it).

In answer to my questions, the associate again returns to the workstation. He taps away at the computer to look up the information we need. He finds the sales price and shares it with me. However, he is unable to locate additional detailed information about the microphone and how it is used. He does offer a warranty and shows me the company's digitized information about warranties. As we discuss possibilities, he admits that, on a personal level, he doesn't usually buy the warranties. He isn't sure if the extra expense is worthwhile in this case. As far as headphones, he calls over a manager from a different department who addresses my concern and offers a few options. It is hard to tell if the manager understands my specific questions on headphones suited for this particular microphone or is just pointing out some new ones that have arrived at the store for my consideration.

After a bit more conversation, I decide to purchase just the microphone, and we head to the checkout. There is a section of self-checkout lanes, with a cashier overseeing the process. "Did you find everything you were looking for?" she asks as I place the item in a bag. I convey that I found the microphone I was looking for but still have a few questions about its use, headphones to accompany it, and the value of a warranty. She gives me a blank stare; perhaps she wasn't expecting such a detailed answer. She mumbles a few phrases, finishing with, "Better luck next time." She waves us goodbye, and you and I are on our way out the door.

Was the shopping trip a success? Well, you might argue that, on the one hand, it was. The store had put in some digital tools, such as products listed on a website and online marketing. There was a digitized inventory system in operation, the information on warranties was available in digital form, and there were self-checkouts in place. All of these are ways that technology had been implemented.

In terms of people and assistance, once inside the store, I found the microphone I was looking for, and a store associate even helped me locate it, so there was potentially an added level of service. Upon further analyzing, I might point out a string of issues that could be warning signs in some cases. Often for retailers, these are closer to flashing red lights of imminent danger, especially in terms of profit and employee satisfaction.

Allow me to expand on these. For our discussion, and to open our minds to envision new possibilities, I'll pose these as questions. First and foremost, in our shopping experience, how do you think the workers viewed their jobs? Do you believe their roles provided them with a sense of satisfaction and purpose, like they were making a difference and offering a meaningful service? Let's also touch on sales: Were there missed opportunities? What could have happened if I were able to better understand how to use the microphone and shown three pieces of equipment that are often sold with it? What about the chance to try out the microphone right in the store? What if the cashier in the self-checkout station would have stepped away from the register and scanning machines and led me through the store to find answers to my questions? What if she would have directed me to an Alexa nearby that could respond to my inquiries? Or put me on a video call with another assistant, either in the store or working remotely?

If you and I went out for a coffee together after our shopping experience and spent our time reviewing these questions, I believe we would end up with more than just empty latte cups. Say we put our heads together and lingered over the brew. We might imagine how the technology retail store could be rearranged, what different responsibilities the employees could be assigned, and how technology and staff could be aligned to enhance the working environment and improve sales. Our discoveries could be so exciting, and so invigorating, that I wouldn't be surprised if, after completing our assessment on the technology retail store, we started to see the coffee shop in a new way and carry out a similar analysis on the workers and brew selection process on our way out.

That's the kind of thrill that digital transformation can bring, but as we observed in our example of the technology retailer, it's not just about providing the digital tools and merchandise. There's so much more that needs to be in place to truly transform, and people are at the heart of this change. After all, they are the pulse that guides our organizations and makes them run smoothly. It's the people who have the potential to move the organization to the next level. Without our employees on board, our digital solutions will have—at best—half-hearted results. With their buy-in, however, the world of possibilities and ongoing innovation opens, provided the working environment is designed the right way.

Let's spend some time looking at ways to adapt a growth-minded atmosphere when it comes to hiring and retaining staff. To give this topic full diligence, we need to consider what today's workforce is looking for, how we

> **It's the people who have the potential to move the organization to the next level.**

can create an organization that sells itself to new team members, and what can be done to weave the threads of innovation into the fabric of our organizations. As we review these factors, we'll see that it's really about creating and maintaining the right mindset ... one that fully supports and propels digital transformation.

Understanding What Employees Want

During the last decades, the rollout of new and more advanced technologies has been changing the way we work and shop. Perhaps before the pandemic, these shifts were often felt and seen in a more subtle, evolving way. A retailer might have added several self-checkout lanes, for instance, and placed them next to a longer set of full checkout lanes with cashiers at the ready. Or a newly integrated inventory system may have been set up, which could tell workers exactly how many items were in stock of a particular product and when to expect more (or what nearby stores carried the merchandise).

When the pandemic hit, these trends accelerated—and in some cases, catapulted—into a new era. Seemingly overnight, workers were sent home and digital shopping transactions were arranged. Customers started tapping in an order and pulling up to the curb to receive it. Other times, consumers may have decided to stay home to make their purchases online and wait for the goods to arrive on their front porch.

The fluctuations of the virus and its global impact gave many, including the workforce, time to reflect. After months of shutdowns, retailers began opening their doors again. Even then, everything didn't revert to prepandemic ways. The technologies had changed (and they continue to change). Shoppers also had new expectations. Employees considered what they really wanted and how they preferred to work. Many started redefining what work-life balance meant for them. As a result, workers have become keener on what they can contribute to an organization. They are more interested in how a company operates and what it contributes to the community around it.

All these happenings have led to a series of shifts, including the Great Resignation, which consisted of more than 3.98 million workers in the US leaving their jobs each month during 2021, marking the highest average on record.[6] Their departures were fueled, in part, by the question of pay. Among those who quit in 2021, 63 percent indicated the pay was too low. With a tighter job market leading to fewer workers for companies to hire, wages have continued to be a main discussion point in the hiring process.

If we dig a little deeper, we see that it's not just about the paycheck. There's another layer to consider on top of wages. It revolves around growth and feeling valued. Sixty-six percent of employees who left their jobs in 2021 stated there were no opportunities for advancement, and 57 percent said they felt disrespected at work.[7]

By and large, workers at every level of an organization want to carry out duties that they perceive as meaningful, fulfilling, and con-

tributing to an organization's mission. Employees are like
sense of accomplishment when they can visibly see the connec.
between their daily responsibilities and the outcome for the firm.
Adam Grant, a professor at the Wharton School of the University of
Pennsylvania, carried out research on this very topic. He found that
team members who interact with satisfied customers for a minimum
of five minutes were reminded of their purpose at work. As a result,
their overall job performance improved.[8]

USE CASE: WORKERS WITH A PURPOSE

Let's recall our beginning scenario, in which you and I go to a store to look for a microphone I want to buy. What if we were to rewrite our interaction with the store associates? What changes would you make to create a scene that was full of purpose-driven employees who felt valued and respected? Perhaps the employee looking for my requested microphone could have access to more information about how it worked. Taking it a step further, maybe he could use some of his work time to test out products. He might try out the microphone himself. He could even help other customers test it. He could then write up lists of suggestions regarding which products would be good ones to cross-sell or upsell with the microphone. He might receive training on warranties and be walked through what-if scenarios on how to educate and assist shoppers. After carrying out these different tasks—product testing, pairing like products through cross-selling and upselling, and receiving training on warranties and educating shoppers—the store associate might be asked if he would like to specialize. He might opt to focus on a certain area of customer service that would line up with his interests.

For the manager, a similar exercise could be carried out. The process could be tailored to the supervising roles of the position. Perhaps the manager could provide customer service training or oversee how workers are assisting customers throughout their shift. If there is downtime in the store, the supervisor might find ways for employees to help online shoppers. The manager could also ask employees to learn more about the store's brand and mission.

When it comes to cashiers, the list of possibilities is long, but it starts with asking the following question: Is the cashier in the best-suited role? If self-checkout stations can be set up, only one or two cashiers may be needed to help customers with questions that come up as they pay on their own. In this type of setting, the cashiers who are no longer needed at the registers could be repositioned to serve a different role in the store. Again, here they could go through an evaluation process to learn what is meaningful for them and how they can contribute to the store's success. For those who remain at the self-checkout stations, their roles could be adapted as well. They might receive additional training on customer service. They could be advised on how to handle customer questions in ways that lead to satisfied shoppers and increased sales. Perhaps they offer customers food and snack items on their way out or point them to the store's café near the exit. Maybe they carry out individualized surveys to gather feedback from customers postsale.

Any time there is a chance to remove repetitive duties for employees and replace them with more meaningful tasks, we are looking at an opportunity to uplevel the working environment. Technology can be used for many of these continuous roles that often lead to boredom for humans, including self-checkouts and updating inventory levels. This leaves room for workers to engage at a higher level, which often means interacting more with customers. Some workers will find this incredibly fulfilling. They will thrive on discovering ways to help improve people's lives, just by coming to work. Others might pursue the concept of creating and bringing new ideas to the forefront. These creative individuals will often find meaning in tasks that give them freedom to think of improvements for processes and products. Providing this type of flexibility for workers reduces workplace tension and increases job satisfaction.[9]

Preparing the Pipeline

When we have workers who feel valued and respected, and when we give them the chance to grow and engage in purpose-driven work, the benefits can create a domino effect for the organization. We might see a company increase its sales year after year. The firm may generate new processes and products and become more efficient. A company that began as a middle player in its industry could rise to take on a leadership position. And leaders could have a better chance of keeping that top spot.

Simply hiring employees and promising them a positive environment doesn't guarantee these outcomes. We must look for the right type of people, and once we have them as employees, we need to help ensure they stay—at least long enough to drive value into the company. If they remain for an extended period, the benefits can continue to unfold. The issue lies in attracting these workers. How do we bring in the right people and keep them?

Talent management, in which we go after people who will be a great fit and who have the kind of mindset we want, starts with the recruiting phase. When we think of gathering talent, let's shift away from strategies that might be considered traditional, such as paging through applications by hand or having résumés sorted digitally. Instead, the process really begins within the organization's walls (both physical and digital). We need to sell our work environment to a potential employee. We must give them reasons to want to join.

When it comes to selling, we have to provide something interesting. If we don't, we risk bringing in talent that is equally as lackluster.

We need to create and play out the mindset we want to have in our company and let it be known and apparent to others. When we do this, our efforts will attract people who are actively looking for the same type of mindset.

A developing trend among the workforce lines up with this concept of creating the right mindset, implementing it, and then bringing in talent to participate within the ecosystem. Job seekers in today's marketplace tend to look for role mobility opportunities, meaning they want to fit their skills into a company's projects and goals. If they can show their skills and ability to adapt, they might expect a next step to be a promotion within the company.[10] Their roles could shift to fit the new project they are given within the organization.

In addition, we must be careful to not overlook the number of jobs people hold—or want to hold. One in five Americans would be interested in having more than one job with one employer, and they seek the freedom, exposure, and flexibility that this arrangement provides.[11] This aligns with another trend, frequently referred to as the gig economy. A worker might sign up for a part-time job at a retail store and drive Uber for several hours during the days they are not present in the store. That same worker might hop online and sell a few items over eBay on the weekend. Their side business might consist of finding, purchasing, and then reselling vintage collections.

At first glance, it may seem that the gig economy and workers' preference to match their skills with specific roles are threats to retailers. Rather than gaining longtime employees, retailers could fear that their workers will bounce in and out, grabbing another gig when the opportunity arises. The reality is that this playing field creates a sense of healthy competition. Companies with the right mindset could come out ahead. Envision a worker who joins an apparel company because she's interested in the clothing sold. She likes interacting with customers and wants the employee discount given for attire sold in the branch where she works. In addition to her store duties, she is a freelance graphic designer and has a handful of clients.

If, after coming on as a sales associate in the clothing store and helping shoppers make selections, she finds the environment to be supportive, she might look to stay. She may pursue ways to grow and change roles there too. What if she starts showing customers ways to properly wear their clothing selections? She could sell more accessories to complete an outfit. She might be given the chance to curate clothing options for repeat customers and advise clients through their entire journey. After she feels comfortable with that, she realizes there are additional designs and clothing pieces she'd like to offer shoppers. Where does she get these? Perhaps an executive from the company hears of this and agrees to offer her a promotion to design clothing. Now, if the mindset is right and this worker loves her job, she might leave her freelancing work behind and become more committed to this top-notch retailer. Or she may continue with the freelancing work and simply turn into a valued team player for an extended time with the apparel company.

There's one piece missing from this scenario that must be addressed. The issue consists of pay. If this same worker is offered a significantly better salary from her freelance design clients, this will be factored into her assessment of the jobs and could impact her final decision. Statistically, she is more likely to choose the job with the higher pay. We saw earlier that many workers have left their places of employment due to what they perceived as low wages.

When it comes to job satisfaction, compensation places in the top five contributing factors for employees, falling just under respect and above trust, job security, and opportunities to use abilities.[12]

Given this, we can see that it plays a role in whether an employee will be attracted to a company and whether they will stay. For organizations that offer a wage that is comparable to others on the market, this factor may be minimized. If a company pays 5 or 10 percent below market, workers are likely to notice. Think of a grocery store that pays workers fifteen dollars an hour. It is located next to a fast-food restaurant with a sign offering to pay seventeen dollars an hour. A potential worker may take those differences into account during the job search. What if that spread in pay were even greater? Say the grocery store remains at fifteen dollars an hour while the eatery ups their hourly pay to twenty dollars and then twenty-five dollars an hour? Workers will gravitate toward the establishment with the higher pay.

Certainly, other criteria play a part in determining wages, and companies need to consider profit margins when setting a pay scale. That said, when we appreciate the possibilities that people bring and how they are the true drivers of transformation, it's hard to underestimate the value they can contribute to an organization. Recognizing that offering a higher-than-average wage could aid in retention is a great starting point. Going beyond that and providing additional benefits, such as flexible hours or access to an online workout program, could help further elevate the overall package. If a valued worker is offered a job at a different company that promises a 20 percent increase in salary, chances are that the individual will seriously consider the new position. Wages often take priority over other concerns, and an awareness of this can help companies set the right salaries for attracting and maintaining talent.

Bringing in Innovation

Have you asked your employees for ways to improve their working environment? When this is done in an open and collaborative setting, the end solutions can be surprising. The methodology doesn't have to be high tech; it could be as simple as putting up a bulletin board and having employees pin their ideas up for everyone to see. At a monthly meeting, these new concepts could be laid out on the table or presented to a group via a videoconference. The employees working in the department could discuss the different ideas and decide collectively which ones merit additional investigation. Over time, these very employees could find ways to reduce costs, create efficiencies, and have a working environment they love. At the end of the day, they'll appreciate the chance to have a voice and be heard. Better yet, if they helped make decisions about which ideas are implemented, they'll see and recognize the results of their efforts. While not everyone is creative, being in an atmosphere that fosters innovation tends to be appealing to a wide range of workers, as those who are more process oriented will notice when certain barriers are removed and they can perform their work seamlessly.

It's not just the employees who have ideas: Longtime customers are likely to have a voice that carries weight too. Take the example of frequent flyers. In my case, I travel regularly and have hit the top status on numerous airlines. I often receive a survey asking about my flight once the plane has landed. I always fill out these questionnaires and add extra feedback when there is a box for comments. No one, however, has ever called me to follow up on the suggestions I've made. I can assure you that if an airline representative approached me and asked about my flight experiences and how the airplane conditions could be improved, I definitely have some ideas to share! So do many other travelers who are often on the road. Their opinion may be more relevant than a person who flies once every two years. While the infrequent flyer certainly is important and their experience matters, they won't necessarily have the same insight as someone who has been through delays, earlier-than-expected arrival times, countless meals and snack bags, turbulence, and the occasional cart bump to have a truly holistic perspective. I, for one, know what I would answer when asked a question on how to improve the experience. In business class, I am often handed a bag of flight accessories, ranging from a pair of socks to earbuds, lip balm, and a sleeping mask, to name a few. I rarely use all of these, and yet I know that after the flight ends, the unused items are thrown in the trash to eventually land in a pile of garbage. I see these accessories distributed on nearly every flight, and I think of the additions being made to a landfill, simply by passing on items like lip balm. What if, instead of handing out prepackaged sets, a flight attendant carried a basket of separately packaged flight items down the aisle and gave travelers the chance to take only what they would use? Or if that same basket was placed at the entrance to the plane so that passengers could grab what they needed on their way to their seat? I can assure you of one change either of these solutions would bring: fewer items sent to the landfill. And more likely than not, it would lead to satisfied passengers who have exactly what they need and are happy about being given the chance to customize their flight experience.

USE CASE: SEPHORA'S TEAM OF ADVISORS IN THE DIGITAL WORLD

When the pandemic hit, shoppers scurried home and ordered beauty products from the safety of their couches. There was just one problem: How could they know if the shade of foundation they were purchasing was a perfect match? If they hadn't ordered the same product in the past, or if they wanted to try a new item, the digital transaction made it extremely difficult to know how it would look on their face. Worse, once the item arrived, it typically couldn't be returned after being removed from its packaging, opened, and tried on, regardless of whether the consumer wanted to keep it and pay for it.

Enter Sephora, with a quick pivot and transformation that opened doors for longtime customers while simultaneously attracting new makeup purchasers. The French multinational retailer of personal care and beauty products has successfully created an online experience that draws consumers in, asking them if they'd like to "try on" makeup via the internet. Once the consumer agrees, they are given the chance to add their photo to the Sephora site or app.

They can then "see" the products they are considering in real time and determine whether that foundation they've been eyeing really is the right tone for them and brings the glow they want.

The Sephora experience for customers, even as shops reopened, didn't go away. Sephora still offers these self-guided, free digital consultations, along with suggestions for what else to buy and other products to consider. Once consumers set foot into a Sephora store, they are greeted by beauty advisors—essentially, workers who are ready and able to give advice, make recommendations, and create a full circle experience for the shopper. And those workers? They enjoy the chance to help customers and be consultants while letting technology take care of the more repetitive tasks like checkout and inventory monitoring.

PUTTING IT INTO ACTION

To fully incorporate people and make your digital transformation a success, take a moment to ponder the following questions:

- What are our current hiring and retention strategies?
- What roles do our workers currently serve?
- How could we provide more meaningful experiences for our workers?
- What type of mindset do we want in our organization?
- What steps can we take to adapt to this mindset and really live it?
- How can we help foster an environment that is open to ideas and encourages personal growth?
- How do our wages compare to others in the industry? What changes could be made to our benefits?

CHAPTER 3

Results, Not Presence

It's important to understand the wide range of possibilities in digital transformation. People can add value to the process, and their deliverables can be increasingly effective—regardless of where they are located and the hours they log in—when we reimagine the way we work.

"My boss has announced that while we're all working from home, the entire company will now be spending the whole workday on a Zoom call with video," wrote in one distressed worker to the "Ask a Boss" section of *New York* magazine.[13] The employee went on to explain that this was created with the intent of making everyone feel like they were in an office setting. Through Zoom, they could ask questions and interact with coworkers. "I can think of no world where this is helpful or anything but highly distracting … I find this demoralizing to the point that I've started job hunting," the letter writer concluded.

The employee was hardly alone, both in terms of Zoom meeting requirements and frustrated employees. As companies have shifted to remote environments and implemented supplementing technologies, a different reality has taken shape. It's natural that moving into this new paradigm has been a bumpy ride for many. In some cases, the shift has created a tumultuous turn of events that leaves managers scrambling and workers disheartened. Take the case of this poor Zoom-ridden worker. The company had invested in remote equipment and video tools, but the use of them was not enlightening. The employee became so unhappy that employment in a different place seemed like a needed outlet. We could also make the point that working hours were restricted. Team members were asked to log in at certain times during the day, as opposed to being given the flexibility to decide when to work.

It's this sense of freedom that today's employees are craving, and getting, when they work in environments that correctly pair technologies and automated solutions with people who can add value to the processes. Making this happen isn't as simple as digitizing processes and equipping remote workers with the tools they need to communicate. It requires us to delve into a new state of mind, one in which we reenvision how work can take place and how employees are treated. It involves moving away from a "presence" perspective and looking at results. We must ask questions to see what workers value, how their outcomes can be tracked, and how technology can support a thriving organization's innovative, transcending mindset.

When Remote Is Right

While the pandemic certainly stirred a frenzy of digital solutions and transactions, work and commerce had begun drifting online long before the first COVID-19 case was announced. Certainly, more workers are plugging in from their homes these days, but let's be careful not to exaggerate the trend. Among companies, 16 percent in the world operate on a fully remote basis. However, 44 percent of organizations don't allow their employees to work from afar.[14] Though more than half of workers log in remotely at least on occasion, it's key to recognize that some jobs still require in-person staff for processes.

Or do they? Perhaps. Let's explore this thought. We could list out some positions that function best when a human is present. A restaurant provides an eating experience that requires cooks, waitstaff, and a cleanup crew, to name a few. An auto dealer gives car shoppers the opportunity to get behind the wheel and test drive a vehicle before making a purchase. An amusement park filled with roller coasters will need some sort of supervision and management to guide riders on their mini adventures. Someone must be present to ensure certain safety measures, such as physically checking straps to verify that they are correctly in place before a ride begins.

All these examples have one factor in common: They are providing an experience for consumers, and this typically involves at least some level of personal interaction. At the same time, there are plenty of tasks that might be automated and not require a person to be on location, even if today those workers are still commuting and clocking in to carry out the task. Take an expert in auto engine components, for instance. Does this person need to be on location? Maybe, if the position calls for manually repairing vehicles. If the role is to consult or offer advice, that same expert might be able to log in from one hundred miles away (or anywhere with a high-speed internet connection) and share wisdom with those who are looking for it.

Deciding when workers should be on-site and when they should be remote begins with an exercise to evaluate current conditions … and then ask, "What if?" In some cases, it can be helpful to bring in an external consultant to challenge the status quo. The goal here isn't to create an uncomfortable environment in which change is felt throughout the organization (as change tends to be met with pushback or feelings of uneasiness); rather, it is about having a fresh set of eyes come in and view the organization from the outside. This different perspective can help dream up new solutions to current issues and decide how technology can be used to support a valuable, high-functioning work crew.

Let's circle back to our beginning example, in which a company switched from working in person to working via videoconferencing channels with constant monitoring of employees. What if an outside consultant came on board? Perhaps the advisor would have spotted other issues at play in the company. The consultant might notice managers who feared losing control and didn't know how to delegate tasks and give employees the chance to show responsible behavior and on-time deliverables. The same expert may have helped the company rethink how they want employee schedules to operate. Would it be best for all workers to be logged in during certain hours for meetings, perhaps from 9:00 a.m. to 12:00 p.m. ET every day? Then they could have the freedom to choose their other working hours, as long as their tasks were completed on time and projects advanced as planned. What if instead of using videoconferencing all day, every day, employees had a ten-minute meeting to start or end the day, to touch base and address any immediate concerns? Better yet, what if that short discussion took place with everyone standing, to encourage a bit of movement?

The outside consultant could also directly ask employees what they expected in their current working environment and what challenges they faced. From there, a plan to solve those issues could be drawn up and set into motion. And the technology? Well, that could be put in to help support the overarching digital transformation.

USE CASE: RETAIL AND REMOTE

How does the work-from-home trend intersect with physical store locations? Do retailers need to shut their doors, move all merchandise online, and keep employees at home? Will high streets, typically lined with locale after locale of renowned brands, fade into the past?

Perhaps the key to success (and avoiding bankruptcy!) lies in a careful study of the landscape. In a later chapter, we'll take a deep dive into the purpose of a physical store. For this discussion, let's center on how employees are comfortable working remotely and, more often than not, asking for this flexibility. At the same time, staff in many sectors, and especially in retail, often are attracted to the job because they appreciate the in-person interaction. Much of the thrill of fashion, for instance, takes place within a store, as a customer tries on an outfit and a sales associate offers recommendations. What size is best? How should this jacket be worn? What colors are most flattering on this individual? How can you make a few years (or pounds) disappear with the right cut?

The work-from-home versus in-store debate may be resolved according to what fits naturally for the retailer and its employees. It might be necessary to keep stores physically staffed, though the number of employees who are present at certain times could be reduced. Times of low traffic and automated solutions like self-checkout can play into these lower numbers. There may be opportunities to shift certain roles to a remote setup. Training days could be carried out via videoconferencing. Online assistance could be given from anywhere. Connecting with customers to conduct surveys also can take place from a distance.

An evaluation of how employees currently work could be undertaken. Those same workers could then be asked to share ideas about which tasks could be done from anywhere. The answers could be compiled into a list of options or a report. Based on this analysis, a hybrid solution could then be drawn up, with allowances for remote work for some roles and positions and in-store requirements as needed. Corporate workers may have different situations than in-store staff. Warehouse employees might have their own set of requests, such as on-site but flexible hours. In many instances, a bit of both—online and in person—provides the chance for workers to flow into the roles that line up with their skills and location preferences.

Work-Life Balance Becomes Work-Life Effectiveness

What comes first for workers, life or work? And is the balance ever perfect between the two? "For several generations, we've organized our lives around our work," wrote Adam Grant, a professor at the Wharton School, in a *Wall Street Journal* article. "Our jobs have determined where we make our homes, when we see our families, and what we can squeeze in during our downtime. It might be time to start planning our work around our lives."[15]

Ask around, and you'll likely find employees already doing exactly this. Remote work has given many employees the chance to carefully select the hours they want to work. More importantly, by definition, remote tasks are not location specific. I had one remote employee choose to travel with his family to Mexico for a period. While he took a few days of personal time during the stay, he then continued working while the family enjoyed the beach and other amenities. In his case, the work that needed to be done was completed, and he had the freedom to choose where and when those tasks could be carried out.

When employees put their lives first, reduced outcomes don't have to be a consequence of the arrangement. Workers who are given the chance to set their own schedules and prioritize their well-being are likely to be more productive on the job. The results can be measured on a project or task basis. Are they carrying out assigned duties? Has their performance improved? Are they up for a promotion? In other words, the career path can continue these days, from anywhere, and in the way that workers choose for it to happen.

THE WONDERS OF AUTOMATION IN REMOTE WORK

As more workers spend all or portions of their time tapping in remotely to work, the relationship they have with their manager has grown in significance. Oftentimes, these supervisors are the primary connector between the employee and the organization. Managers are also among the first to hear concerns about working conditions and other job-related complaints.

> **Workers who are given the chance to set their own schedules and prioritize their well-being are likely to be more productive on the job.**

Thanks to technology tools, managers don't need to spend ample time carrying out routine tasks. Scheduling, project management, and approving expense reports can all be done in a more automated way. Advancements continue to roll out, freeing up more managerial hours. It is estimated that by 2025, 65 percent of managers' current duties will be automated.[16] Essentially, supervisors' responsibilities are moving away from hands-on, detailed, and repetitive tasks and flowing into soft skills like relationship building and qualitative contributions.

Facing these trends, companies have the opportunity to rethink managers and remote work. How many managers are needed? What roles should they carry out? How can they build connections with workers they see only via video or through Zoom and occasionally on-site? How can supervisors shift from managing tasks to managing the employee experience? An evaluation might lead to managerial responsibilities that include looking out for the career trajectories of employees, spotting red flags if work/life issues become out of balance, and encouraging the bond between the worker and the company. A manager's role could even include coordinating in-person events, lining up professional speaking events for remote workers to see, or helping to walk employees through coaching apps.

A PROTEST AGAINST SITTING

Working from home means a chance at better health—or is it the other way around? While some employees are finding that x-ing out the commute gives them more time to exercise, others say they have become more sedentary.[17] The same is true for remote working and nutrition. The ability to prepare a lunch right in the kitchen a few steps away from the home office means workers can potentially access healthy meals for dinner. That's not always the case, as it's also easier than ever to order takeout, purchase processed foods, and consume large portions right in the comfort of a living room while watching favorite shows. In the United States, around 38 percent of men and 40 percent of women are considered obese.[18] In terms of the employee pool, this translates to approximately four out of every ten workers who are overweight and at increased risk of type 2 diabetes, high blood pressure, heart disease, and certain types of cancer, among other diseases and health conditions.

What does this mean for companies and the roaming workplace? Should organizations encourage well-being and healthy lifestyles? Workers who rank high in terms of well-being generally are more productive, which is beneficial for the teams in which they participate. Healthy workers also have fewer medical absences and lower costs associated with doctor visits and treatments.

Setting up programs to support good health and well-being is one route companies are taking.[19] The best-laid plans, however, will only lead to results if they are implemented and there is engagement among employees. Another strategy to consider might involve having mentors available and in touch with workers.

> **Workers who rank high in terms of well-being generally are more productive, which is beneficial for the teams in which they participate.**

This can be effective if individuals are lined up with their passions and experiences. For instance, an employee who has lost forty pounds might be interested in becoming a mentor and encouraging five to ten peers or reports to reach their own health goals. Coaches can also be brought in to provide support and guidance on employees' well-being journeys. The process starts with an evaluation of what exists and what employees are seeking. A survey could be sent out asking workers, "How satisfied are you with your remote working arrangement? How could your home office be improved? How would you rate your health and well-being? What are you interested in improving? How can we help?" Another option: have managers check in periodically, such as every week or once a month, to see how employees are doing on a physical and emotional level.

Revenue Opportunities from Remote Work

What does moving to off-site environments look like in terms of cost? There could be some initial savings. For instance, if office space doesn't need to be rented, some previous costs could be eliminated. However, there will be investments that are made for technology. The driving factor of technology implementation should begin with conversations about how employees will work and what sort of transformation the company is undergoing. Do software programs, security solutions, and at-home equipment for workers to carry out tasks and stay connected mean another line on the expense section?

Not necessarily. In fact, the opposite can be true. Giving employees the benefits we've discussed, including freedom and flexibility, along with the right pay, can lead to satisfied workers who contribute, innovate, and stay with a firm. Furthermore, providing the environment and the mindset where new ideas can bubble up, with the best ones rising to the top and being adopted, will produce efficiencies. In addition to cost savings, there could be opportunities to build new revenue streams and increase profits.

Take the case of employees who work online and focus on helping website shoppers. These individuals could have access to data that identifies where the customer is on their journey. Have they made purchases before? Do they typically buy the same things? How much have they spent in the past? What might they be interested in seeing, based on their previous selections? The workers could then reach out and assist these shoppers via video streams, which offer a more personalized look and experience. The results could be that more merchandise is sold, simply because customers received help finding what they were looking for and getting their questions answered.

The focus here doesn't have to be on serving high-end customers or creating better experiences for the luxury market. There are opportunities in every sector, from grocery stores with workers who suggest sale items, accompanying items, or extra services such as delivery, to fine jewelry, with employees who can listen to shoppers' situations and ask about upcoming events that might require new accessories.

TURNING A CALL CENTER FROM A COST TO A PROFIT DRIVER

From a business standpoint, call centers are generally necessary—and thus a required expense on the balance sheet. Customers need a place to turn for help with their orders, returns, questions on products, and warranty issues. The costs for call centers typically include wages for the workers, equipment to take calls and manage customer data, and other overhead expenses such as buildings or leases. When outsourced, call centers still have a price tag attached. In today's remote working environment, there are costs associated with labor and equipment, even if other overhead and location-specific expenses are reduced or eliminated.

What if I were to tell you that a call center could be flipped from a cost center to a profit generator? The possibility is there; we merely need to rethink our approach to customer service. One way to do this involves meeting customers at the pain point of wait time when they call in. In some cases, those who made a purchase may have to hold for ten minutes, an hour, or more to be able to speak to a representative. Remember that when they place the call, they are connecting for free. What if they were offered the chance to be served immediately ... for a price? Or if they were given an opportunity to pay for an agreement that their call would be taken within ten minutes? Perhaps even thirty minutes? As we can see, there are opportunities to provide a service in the form of convenience. If different price points are given, consumers can decide for themselves if they would like to wait for free, pay to be served soon, or pay a bit more to connect right away.

Following this scenario, we can envision the framework of a revenue machine. Callers pay to be served quickly. There may be additional staff required to assist customers. However, the price points could be adjusted so that more comes in from customers than goes out to cover expenses. As an added benefit, customers may be more responsive and less agitated when the company representative comes on the line before they have been forced to wait for two hours. And if they did wait for two hours, they might feel they were saving money. This in turn could also lead to a more agreeable state. The better attitude among customers can boost the experience for the company representatives. The call center itself could gain a higher reputation in the company, as its workers would see the value they are providing for the company.

A profit-generating call center may work well for some niches. Others might find that customers aren't interested in paying to save time. Organizations could run a trial version for a short period, such as several months, to see if it's a good fit.

USE CASE: PRIMARK SHIFTS TO HYBRID WORK

Under the motto Your Day, Your Way, Primark, the multinational fast fashion retailer with headquarters in Dublin, Ireland, gives its office employees the freedom to plan their working day in a way that best suits them. The corporate core hours are set from 9:30 a.m. to 4:30 p.m., with an hour break for lunch, from Monday to Thursday. Friday starts at 9:30 a.m. and ends early, at 2:00 p.m. While meetings typically take place during these times, staff can choose when they want to start their workday and when it will end. In addition, employees can opt to work at one of the Primark office locations, a designated store, depot, or their own home. Workers note that the flexible arrangements increase their productivity, improve their well-being, and allow them to make the most of their family and leisure commitments.[20]

When we discuss remote and hybrid workplaces, it's important to recognize there is no one-size-fits-all solution. There's no need to expect that all workers, everywhere, will be tapping in from their homes, as we've seen how experiences and in-person interaction still have a strong foothold in the world of commerce. At the same time, we must recognize changing worker expectations and preferences. We will need to meet those in a way that is both productive and profitable for our organizations. When we open our minds to rethink old ways and explore the possibilities of what if, new discoveries will unfold.

PUTTING IT INTO ACTION

To focus on results and increase profitability in a flexible working environment, spend some time evaluating the following points:

O How and where do our employees currently work?

O What do our workers want in terms of life-work integration?

O How can we reimagine in-person, remote, and hybrid work environments for our company?

O How can we focus on results in different working environments?

O What are some ways to build new revenue streams from remote workers?

O What do our managers and leaders need to care for employees in different settings?

CHAPTER 4

Reimagining the Customer Experience

The customer is at the center of everything.

During my travels, not long ago I landed at a hotel in Germany. I had stayed at this place on previous trips; this time, as before, I made a reservation for the night. When I entered the room, I found that the temperature was set to the same degree that I had turned it on my last visit. I clicked on the television, and the images reflected the programs I had watched during the downtime of my prior stay. Obviously, the hotel had kept track of my preferences and adjusted the settings so I could have a personalized room when I walked through the door.

This hotel didn't fall into the category of luxurious; in its industry, it might be considered to have clientele who were more mainstream, and yet the place was catering rooms to the exact likes of its guests. In a way, it had found a method to uplevel the experience and place the guest first. There was a high-end feel to its amenities. The key takeaway for me from this experience lay in the fact that the hotel had systems in place not only to gather data but also to put it to use for repeat lodgers. Chances are, if a guest returned after an initial visit and found these customized settings, they might be drawn to come back a third time, and a fourth, and so on. What if they advertised the personalized experiences to first-time guests? That might be an attraction for travelers looking for a new place to stay.

Without good customer service, there's really no way to make any experience, regardless of the final price tag, feel high end. Let's also note that putting the customer first isn't just good business practice; it's expected among today's shoppers. More than 80 percent of consumers pay as much attention to how brands treat them as they do to the product or service being sold.[21] Best of all, there can be a payoff. Personalization can reap a 300 percent return on investment. In some cases, the return is even higher.[22]

> Putting the customer first isn't just good business practice; it's expected among today's shoppers.

When we revolve our practice around the customer, we strive to better understand who these segments are. We delve into their journeys to discover what pain points they have. We also will want to evaluate how our current systems operate. We'll be inclined to look for ways to listen to customers to gain insight and innovate accordingly. It's about transforming how we view and respond to customers and not merely putting in digital processes without a clear purpose for them.

Understanding Our Customers

In today's intricate world of commerce, it's often not possible to walk through a store and identify all the customer segments by surveying who is there. Take the case of a pharmacy. Consumers might stop by every month to pick up a prescription. Some could order the prescription online on occasion and have it delivered to their home. They might come in the store every week to purchase personal care products, or they may never walk through the doors and have all items sent to their home. While some clientele may live nearby, others could be passing through on a trip. Caregiving adults might stop by to shop for their aging parents. Still others live farther away but tap into the website to buy the items they want. A single pharmacy location could serve customers interested in one or more of the following: medications, vitamins, health products, senior gear, souvenirs, photos, beauty supplies, fragrances, gifts, seasonal decorations, and anything else the store carries. Since these consumers could be in store or online (or both!), a mere glance inside the walls of a building won't determine all the types of buyers.

In a certain sense, there are established methods to gain a grasp on our customer base. We can draw up personas and use these to describe our target audience. We can look through surveys to identify recurring responses and customer needs. We might even sift through data to see details about who is purchasing, where they live, and what they typically buy.

While these strategies may be helpful, they also carry limitations that are important to recognize. For instance, how are the personas created? Are they made in a way that accurately describes shoppers in today's changing landscape? What about the type of surveys? If they are confirmation based, such as asking customers to verify that they received a great service, they could be missing out on their potential. Buyers might have more feedback to share. If all they are given is a survey that asks for a one- to five-star rating, they won't be able to explain their suggestions. When it comes to data, there is no lack of it floating around. The question centers on how it is being applied. Among executives, 63 percent feel their companies aren't making good use of their analytics.[23]

WALKING IN CUSTOMERS' SHOES

Sometimes there's no substitute for putting ourselves right where we want our focus: the customer. Immersing ourselves as a customer for our own company provides the chance to witness so many different components of the organization from a fresh perspective. Think of it as moving away from the view in a plane flying at thirty-five thousand feet and descending to ground level. That's where the experience for a customer begins.

Perhaps we are a company that produces building block sets, which are used by children and adults. Let's become a customer for a moment. Where does our journey begin? It will depend on what we want to buy. Maybe we are looking for holiday gifts for our young children. How will we learn about the latest building block sets that are available? How can we find out if they are appropriate for our children and the benefits they bring? Why should we choose these products over others on the saturated toy market? These are all points along the path, which will often take place via internet searches, word of mouth from other parents, influencer recommendations, social media conversations, and our own history. There are also online ads that might reach out and touch us along the way.

After walking through these steps, we decide to go ahead and look at the building block sets online at the company website. How are the categories organized? Which ones are available, what sells the most and why? We'll have to look through descriptions and try to get an accurate idea of what's being sold (is there a video available showing a child putting the set together with a parent? What are the dimensions and color options?) We might read reviews too. And we'll want to know if the items we're considering are available for shipping, the costs involved, and if they will arrive in time for our celebration and gift opening. We might also go to a store nearby to see the toy in person. We could appreciate a reminder after we visit the site that there is merchandise for us to see and think about.

Our journey as a customer doesn't end when we click "buy" or go to a store to get the item. We take the purchase home, wrap it, give it to our child, and then evaluate it as they play with it. Does the difficulty level match up with how easy (or hard) it is to put together? What might be added to the product to make it more appealing to children and families? Did the company reach out to us to see if we were happy with the product? How can we learn about more items? Would it have been helpful to have someone walk us through the whole process, such as an online assistant to jump on a call, chat, or video? In such a case, we might have relayed our situation and asked for help. A bit of guidance could have saved us hours of searching and evaluating. And a follow-up after the transaction could present the opportunity to share feedback and even make another purchase.

THE POWER OF AN EAR

In addition to immersing ourselves in the customer experience, it's hard to overestimate the value of listening to consumers. Imagine this process as a shift away from a standard survey and an opening of a door. It's a way to hear straight from buyers about what they want, how processes could be improved, how an experience could be elevated, and how more sales could be made.

Now there could be a certain level of noise that this type of exercise might bring. For a company with ten thousand customers, it may not be possible to have lengthy conversations with every single shopper immediately after a transaction is made. Or perhaps it doesn't work to ask every visitor who lands on a website to share why they came and what they are looking for.

However, there are channels that can be set up and systems developed to gain valuable feedback. Data on customer transactions could be gathered and analyzed. A segment of employees could review the data to identify repeat buyers. Once a customer meets certain criteria, such as a set number of purchases or an amount spent, the worker could reach out to contact them. This step might involve a conversation or an open-ended survey. And there are ways to get very creative here. What if promotional items were sent to the customer in appreciation of the feedback? Or a product that they purchased in the past shipped to them for free? This give-and-take can create a channel that effectively gathers ideas from customers.

Drawing on our example of the hotel with customized features where I stayed in Germany, this very scenario could have played out. For instance, a hotel employee might have contacted the hotel's most frequent guests and asked how their stay could be improved. A guest might have suggested that customized settings would make it feel more personal and save time when getting settled in a room. I know in my case, after staying at that customized room in Germany, I would have offered an additional recommendation. I would have advised a move away from bars of soap. Instead, a soap dispenser could be installed. This would cut back on waste that accumulates from throwing out individual-use packages and the remaining contents of the soap after the stay.

Once this type of valuable input is obtained from customers, it can go through a validation process. Perhaps an advisory board gathers the ideas and decides which ones to pursue or test out. Those that are carried out in pilot projects and perform well could go on to another step, such as implementation or further review. In some cases, the ideas might not stick, but the generation of options and an evaluation process will generally help the best ones rise to the top.

More Than a Website

When organizations first entered e-commerce, for many brick-and-mortar locations, this consisted of getting a domain. Then came a listing of products, checkout systems, security components, and online advertising. As the internet advanced, additional layers took priority, including search engine-optimized (SEO) content, an emphasis on branding, and a focus on simplicity and ease of use.

As we look around today, we see consumers engaging in more interactive ways in the online space. They're shopping; they're also staying connected to loved ones, getting updated on current events, planning trips, carrying out their work ... in a sense, they are living out a good part of their days online. For this reason, a website that comes across as stagnant or impersonal—essentially, unable to provide an experience—may not attract them in the same way that a website did decades ago during the novelty stage of the internet.

When we start to imagine the ways experiences can unfold online, new possibilities pop up. Could consumers "try on" their clothes by adding a picture of themselves to a site? Could they have some sample products sent to them, based on their engagement with the site? Is there a place where they can ask questions and get relevant, helpful advice? Can they receive messages and notifications after they leave the site that are based on their browsing history? All these are ways we can uplevel that online experience and make a website be so much more.

BEYOND THE SHIPMENTS

Prior to online shopping and sites like Amazon, it wasn't unusual to wait several weeks, and in some cases even longer, for an order to arrive. In the last years, the boom of e-commerce has driven an on-demand expectation. With Amazon today, customers can place orders they receive literally the same day—provided they live in the right location, which is typically an urban setting, and are willing to pay enough to warrant the speedy delivery. These quick drop times have led other companies to follow suit and set up systems to get packages out fast. Supermarkets and services like Instacart promise to deliver orders within a matter of hours, and in some cases, minutes.

Is speed what customers want? Should systems be set up to deliver as fast as possible? Maybe, but I would suggest that we consider other questions along with streamlining logistics. For starters, is speed the top priority for all your customers? Would some find it helpful to have extra support when shopping online? Would a customer segment value real-time assistance from anywhere in the world over a fast delivery?

Rather than promising fast delivery (or in addition to fast delivery), let's say a bike retailer employed an online sales associate to show customers the different colors available for a certain bicycle model. The company representative could also share other interesting data, such as which type of bicycle was currently the best seller and what attracted customers to it. A clerk might listen to a consumer's lifestyle via a video call and then recommend a model that could fit their activity level. An online representative could even check a customer's location and determine whether steep hills and curves will be an issue when biking in their area. The sales associate might explain whether extra gears are needed for tough terrain or describe how a one-speed bike could work well on the flat, smooth roads of a downtown neighborhood. That same clerk could send a personalized message reaching out to online visitors when they leave the site or an online cart, asking if they'd like to return.

USE CASE: JUST DOING IT: NIKE PUTS CUSTOMERS FIRST

When visitors enter the Nike site, they're invited to join as a member—for free. This act gives them access to free shipping and returns as well as exclusive products.[24] Those who sign up receive welcome emails to describe their benefits. These messages also include instructions to download their apps and connect with Nike experts.[25] Members can access the Nike+ rewards program, which gives them early access to new products, priority access to certain events, and personalized workouts.[26]

Through its apps, including Nike Training Club, Nike SNKRS, and the Nike app, the brand collects data about customers. Users can share what sports they like, what sizes they wear, and what colors and designs they find appealing. This information is used to customize the products that shoppers view in their app. It's also channeled into Nike's system to help determine what designs to make and which items should be stocked in their stores. When a customer who is a Nike member walks through the doors of a Nike locale, the company already knows about their interests and can use the insight to create a more personalized on-site shopping experience.[27]

PUTTING IT INTO ACTION

To put the customer at the center of the commerce experience, take a moment or two to think through the following questions:

O What do we know about our customer journeys?

O What systems do we have in place to keep customers at the center?

O How can we reimagine the way customers experience our stores and online locations?

O What ways do we currently use to gather customer feedback?

O Are there opportunities to listen more to our customers?

O How can we personalize every engagement with our customers?

O What can be done postsale to interact with customers?

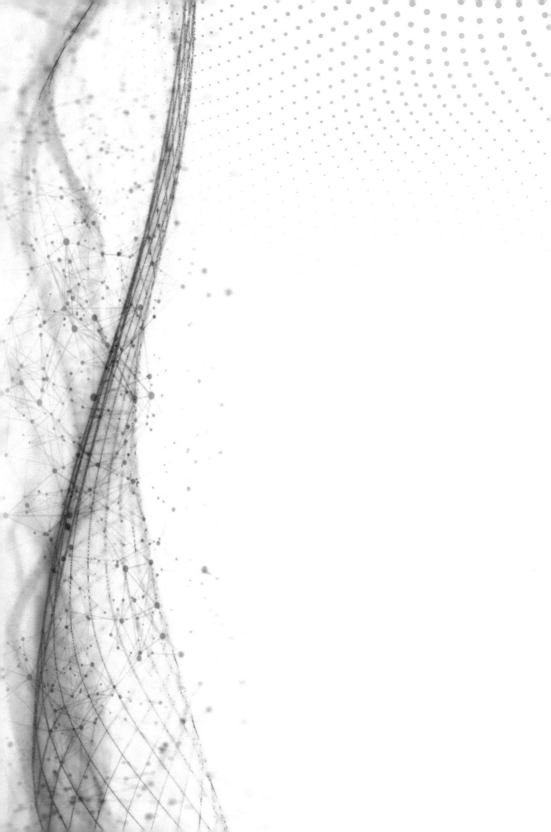

CHAPTER 5

Mobile at the Center of Everything

Phones are intricately attached to customers, and the ways to leverage this connection begin with getting creative.

Have you heard the microchip conspiracy theories circulating about the COVID-19 vaccine? The basic premises of these center on the claim that the vaccinations for the disease contained tiny electronic devices. As the theories go, a person walks in to get injected for what they believe is a protection against COVID-19. When the needle hits their skin, it inserts a microchip that can be used to track the person. After receiving the immunization, the person goes on their way, and the government—or whatever entity is behind the microchip—can follow their every move.[28] During 2021, one in five Americans stated they believed there were devices in the vaccines.[29]

However comical (or not so comical) these theories might seem, I bring them up because they offer a good beginning point for a discussion of technology and how intricately connected we are to certain devices. Interestingly, one of the influencers behind the microchip vaccine conspiracy theory was a retired Mexican cardinal in the Catholic church. Juan Sandoval, the emeritus archbishop of the archdiocese of Guadalajara, posted a video that contained a warning for viewers. He stated, "The chip they are planning to put in the vaccine to control you, it is the mark of the beast." He shared the clip, titled "The Plot to Impose a New World Order without Christ," on, of all places, Facebook.[30]

I don't know about you, but I'm pretty sure if I ran into the retired archbishop, I might mention to him that Facebook tracks quite a bit of information. Government leaders, along with disease specialists, have taken a stand against the theory, insisting it isn't a practical concept. "That's just not possible as far as the size that would be required for that microchip," stated Dr. Matt Laurens, a pediatric infectious disease specialist at the University of Maryland School of Medicine who served as a coinvestigator on trials of some of the COVID-19 vaccines.[31] Facebook, on the other hand, has all sorts of data about its users, including the information you put in when creating a profile (like your birthday), the photos you post, and the updates you share. If you sign on to other sites by using Facebook, which is frequently offered as an option, the company can track what you browse while online.[32] The audience who watched Sandoval's video about the microchip vaccine was likely tracked by Facebook.

It's not just Facebook, of course. Other companies and sites track information and follow visitors. Part of the basis for this stems from the technology itself. We have tools that can be used to gather information that wasn't available in the past, before the development of the computer, the internet, and communications.

In the last ten years, some of the advances have made incredible leaps and bounds. Just think of all that your phone (which is probably either on you right now—or if it's not, you know where it is!) can do. We use our mobile devices to send and receive messages, check our bank account, look for recipes, plan a trip, order food, and make restaurant reservations. The iPhone in today's world has one hundred thousand times more processing power than the computers used for the *Apollo 11*, the first US spacecraft to land astronauts on the moon.[33]

When I heard about the conspiracy theory related to vaccines and microchips, I thought not of what my arm might contain after getting a shot but of our phones. *Don't we already have tracking devices on us?* I wondered. We use our phones so frequently and engage with companies and entities that are indeed tracking us. They know where we are, what we search for online, and how we go about our days. This close interaction means that when we think about commerce, we can't overlook the forces related to our mobile devices. Even if customers aren't thrilled about having their information in public, they are also constantly on their phones, and there are significant benefits that these electronic advancements bring to their lives.

There are also big advantages that mobile can provide for companies and brands as they engage with customers and sell merchandise. Let's look at the need to include mobile when building our customer journeys and the different ways employees can get involved in the process. We'll also consider how to leverage the power of each type of device being used along the way.

Mobile in the Customer Journey

We use our phones in our everyday activities, from watching videos our friends send us to checking Instagram, surfing Pinterest, posting on Twitter, and making transactions. The average American spends nearly three hours on their phone each day. Over the course of a year, they'll be looking at their mobiles for almost a month and a half in all. It's the first thing most people check in the morning, and three out of four feel uneasy if they leave the house and forget to grab their phone. More than half of phone users have never gone more than a day without checking their device. On a usual day, they're tapping into it once every four minutes.[34]

Clearly, we are obsessed! This doesn't mean the phone is an addiction (though of course there could be an argument made for some exceptions, such as users who struggle to ever disconnect or are unproductive due to their phone habits). Rather, it shows the important role this device plays in our minute-to-minute activities. The phone gets used at home, at work, and everywhere in between. It's the way to communicate, interact, live, and, of course, shop.

When is the last time you looked for information about products online? Nearly 70 percent of shoppers search for their merchandise on the internet before making a purchase. Many of these investigations lead to company websites, which serve as an online storefront. Just over 60 percent of online retailers say their traffic comes from mobile devices.[35]

These statistics, coupled with the fact that our phones are so much stronger than early spacecraft, mean that the potential to use mobile devices in commerce is incredible. Most importantly, when we map out the way customers come to organizations and find products, we must not hesitate to include phone use on their journey. Mobile devices could appear in the beginning of a customer path, such as when they first learn about a product or company. The devices could also appear during the middle stage as they evaluate merchandise, at the transaction stage, and even postsale.

Think of how a phone might come into play on your own journey as a customer. Perhaps you want a new pair of sunglasses. You start by unlocking your phone and looking online to see what's trendy this season. You check on a few influencers you follow who regularly post videos on YouTube about their shopping hauls. When you head to the mall for an afternoon to run a few errands, you stop at a kiosk that sells sunglasses and try on a few pairs. You don't buy then, however. While stopping at the food court for a bite to eat, you research prices for a pair that you liked. You find a good deal at a retailer site. You don't want to make the purchase while out and about (you're tapped into the public Wi-Fi from the mall, after all) so you head home. Sitting in your recliner, holding your laptop, you power up the device and head back to the retailer. You bring up the pair of sunglasses you've decided on and click through to purchase. Five days later, you're driving to work in your car, with the new shades in place.

Notice how this path stepped in and out of the phone. It began by a mobile online search, drifted out to on-the-ground engagement, and picked up again for the price search. When we draw up customer flowcharts and graphs, it's key to keep in mind that phones will almost always be involved at some point. How will our website look when viewed from a mobile device? How easy is it to ask questions at our site online? How can we keep in touch with customers after their initial visit? Can they sign up for a deal in exchange for their email address or phone number? Are they open to receiving texts for further engagement? All these questions can help us make sure phones are integrated into the customer journeys.

Employees on Their Phones?
Yes, It Can Be a Good Thing

It's not unusual to see employees at work—with their phones at their side. Now sometimes companies may put restrictions in place, such as asking workers to turn off their phones at work or requesting devices to be deposited into a certain basket or locker during working hours, where they will be retrieved at the end of the shift. By and large, however, if you walk into an office, it's likely you'll see the receptionist at the computer, with their phone to one side. Cashiers might have their device in their back pocket, and warehouse workers may have it handy so that, on their lunch break, they can catch up on social media.

What happens during the downtime? After an employee stocks shelves and waits for their next task, where do they turn? If you're like me, you've spotted employees who work in the airport, supermarket, and retail shops, among other places, browsing their Facebook pages. They're also responding to personal texts. During free time like a break in workflow or a pause before a meeting begins, you're bound to see employees on their devices.

Whenever I see this occur, I think about the following question: Why not have these workers use the phone for work purposes? In exchange for helping customers online, employees could be offered increased pay. Or they might receive a bonus when reaching a certain milestone, such as assisting a set number of customers.

In this arrangement, employees could use their downtime between serving customers to check a company chat line that helps online customers. Through this texting service, customers could write out a message that explains what they are looking for and ask questions. They might not need an immediate answer, so when an employee has a moment, they can hop on the chat line and respond to the customers' inquiries. The shoppers are happy because they didn't have to call and wait to be served, nor did they leave their homes and take the time to go to the physical store location. If they receive a response within twenty-four hours, they might be satisfied and ready to take next steps.

There are other ways employees can contribute to an organization's mission by using their phones. Maybe they could look at the company's social media page and respond to Instagram followers who leave messages. Perhaps they could upload photos to help promote products in the store or advertise an upcoming event. They might hold a live chat, give a presentation, or make videos, all with the brand's theme and vision in mind.

There's more behind this strategy than taking employees off their personal use of phone time and directing it toward company efforts. It involves a mindset that keeps customers and their mobile use front and center. It helps workers understand that shoppers are tapping in via other ways than walking into a brick-and-mortar store. Finally, this blend of mobile and on-the-ground methods is a good fit for today's lifestyle and commerce approach. As we have discussed, phones are everywhere, and if they are being used during the customer journey, we will do well to also tap our employees into them at different touch points.

Keep in mind that those employees always have their phones on them, whether they are at work or working out at the gym. If they have the tools installed on their phones to communicate with customers through chat lines or customer support centers, they'll be able to work from anywhere. Certainly, you might ask them to help online shoppers during their downtime while they are inside the retail space if that's where their regular working hours occur. Thinking about this further, could workers log in to work from other places? What if they put in a few hours over the weekend, when they are at their apartment? The advantages here are great: this setup can be used to expand our working boundaries and support systems.

One final benefit of this approach can come from an uptick in productivity … and a potential increase in wages and sales. If employees aren't so focused on the personal use of their devices and are tuned into the company's mobile presence, they are making the most of their time on the clock. If they have goals or milestones to meet, workers might look for extra compensation when they accomplish their tasks. This means more business for the company, so an increase in wages might be offset by higher sales. Employees will feel empowered to do more, which could lead to improved morale and positive feelings about the company. Workers will see themselves as contributors and team members, all while engaging via mobile devices with customers.

For organizations, the approach could lead to more personal relationships with customers as well as access to the services they are looking for when they are looking for them. This ability to be approachable and available can help move customers through their journey and to the final transactions. Of course, as we've mentioned, the relationships and continuing communication extend to postsale interactions. Workers can check in with customers who have made purchases in the past to ask for feedback. They can also reach out to offer new promotions and raise awareness about upcoming products.

Leveraging the Power of Each Device

As vital as it is to recognize the ongoing connectivity we have with our phones, it's equally important to note that they are just one device of several that are readily available and used daily. Think of the laptop I listed in the example of purchasing a pair of sunglasses. In that transaction, both a cellular device and a computer were used. In addition to desktops, customers might hop on a tablet or even look to their television for information when shopping.

Not all customers will be comfortable making the final purchase on their phone. They might readily tap in to read reviews about a set of high-end binoculars on a cell phone. However, they may be anxious about sharing personal information and carrying out the full transaction. Security concerns are a main reason why some consumers will look to a desktop when it comes time to enter their credit card information.

The size of the text is another motive. On a mobile device, it might be tricky to read small print. Consumers will lean into devices with larger screens to make sure they are selecting the right size, color, and brand before clicking on to a shopping cart. Again, here the shopper might turn to a laptop or their home computer to better view the item and its details.

For this reason, there should be a button that allows customers to transfer what they are looking at, or what's inside their shopping cart, from their mobile device to their desktop. Think of it this way: Rather than asking the customer to find what they want on their cell phone and then move to their laptop and find it all over again, we are providing a way for them to seamlessly continue their experience. With the right tools in place, customers can readily bring up something on their phone. They can then shift over to their desktop to make the purchase. This breaks down barriers along the customer path. It makes it easier to move through the search stage and enter the transaction phase.

Another way to amplify how a product looks and performs lies in leveraging the capability of television screens. Imagine a customer who starts looking for a piano online. They find the dimensions of an instrument that catches their eye. They measure it out in their sitting room, where they plan to play music. What if they could also cast the image of the piano on their television screen? Think about the advantage of size that could be gained here. Rather than looking at a tiny image on their phone screen, or studying a slightly larger display on their computer, they could view a bigger version of the piano. Suddenly they can see the details of the piano keys, the bench that comes with it, and a person delicately making music on the ivories. Again, here thought needs to be given to how customers can move from the television screen to another device and easily continue to browse or make a purchase.

When it comes to leveraging the power of location tools, the phone wins above other devices. Thanks to technology and cell phone networks that make it possible to know where a person is, ads and promotions can be directed with a local slant. What if a person is walking down a street of shops and searches for a jewelry store in the area? If you're the owner of a jewelry retailer located just two blocks away, you'll want that customer to know where you are!

Having your up-to-date information on Google Maps is a great starting point. Then make it easy for browsers to see the hours you are open. You can share what you offer and show images. Better yet, you could have employees available to answer questions—either online or in person.

USE CASE: AN APP TO VISUALIZE ON YOUR PHONE

When you want the perfect dresser for a new bedroom, where do you turn? For many, it's to their phone. More than one-third of Americans have visited Wayfair's online home goods and furniture shop, which prides itself on featuring a "zillion things across all styles and budgets"[36]—i.e., something for everyone.[37] The majority of customers place orders with the company—from their phone.[38]

When the company released its app, it had customers—and their cellular devices—front of mind. "With the Wayfair mobile app, consumers have the flexibility to discover and visualize their favorite products among millions of options, then design their rooms from start to finish, no matter where they are," Matt Zisow, vice president of product management, experience design, and analytics at Wayfair, stated in a press release at the time of the app refresh, which featured tools to improve the mobile shopping experience, in 2019.[39]

One of the app's benefits includes the chance for customers to view furniture options right in their own home. Through the power of augmented reality, the shopper can stand in a room at their house and "see" products in their places to understand how they would look once they arrive. In addition, customers can use the Interactive Photo, which lets them snap a picture of their room. Then they can bring in multiple products to view in the space. However, they don't have to physically be in the room. They might be at the park, on the bus, or at a restaurant. Wherever they are, they can breeze through furniture pieces they are considering on their phone. A big plus is that the app uses technology to understand the dimensions of the room and places the item inside according to its measurements. Customers can get a real sense of how a coffee table might fit in comparison to the other pieces of furniture in the room. Will it be too big and awkward? Is it so small that it will appear out of proportion with the couch and rug? All this can be done without stepping foot inside a store.[40]

With an eye for creativity, the Wayfair app also offers the Room Planner 3D, which gives shoppers the chance to create an interactive 3D room—again, right on their phone! They can then play with different options. What about a new dining room table? Which lighting fixture will look best over it? Will the chairs fit properly in the space? What extra touches can the room have? Which colors go together for a modern look? This mix-and-match arrangement is much easier than hauling different tables to the dining room to view them![41]

All these mobile features help solve the age-old puzzles that so often surround furniture shopping. Customers no longer need to measure a room, bring a tape measure to the store, buy a new piece, and hope for the best. They don't even have to view dimensions presented in a static way on a site and make an online purchase, figuring they'll return the item if it doesn't come as expected. Instead, shoppers get to design, imagine, create, and view options in an interactive, personal way. By the time they fill their shopping cart and hit "buy," they can be confident they're getting just what they need.

Indeed, mobile is interlaced into our daily lives. Phones are nearly a part of our very selves. They may not place a chip in us (like the COVID-19 vaccine conspiracy theories presumed), but they are deeply woven into how we behave every day. This trajectory of continuous cell phone use will persist in the coming years. Technology, which has advanced to the extent that the computers used to land on the moon now seem basic, will continue to add power to phones and their capabilities.

These trends don't have to leave us feeling negative and wondering when the cyborgs—those infamous sci-fi creatures that are part human, part computer—will appear at our doors. Instead, companies can leverage the potential that mobile experiences can provide. Changing our mindset to consider them in the customer journey, encouraging employees to tap into their phones for work purposes, and making the most of the different benefits each type of device offers is a great start.

I encourage you to spend some time thinking about this discussion and applying it to your own areas of duty and organization. What if we took a pie-in-the-sky approach? Let's presume there are no limits. You have an endless budget and unlimited resources for personnel. What would you do? List it all out; once you review it and make cuts to accommodate for reality, chances are you'll still be ahead of the game. You'll have better relationships with customers who literally live with their phones at their side. Those consumers will want to reach out to your organization for what they need. This can lead to more transactions and improved connections.

PUTTING IT INTO ACTION

To get into the right mindset when it comes to the mobile experience, take some time to sit with others on your team and sort through the following questions:

- What is our current approach to mobile use?

- Do we regularly connect with customers via mobile experiences?

- What are shoppers looking for when they come to our site on their mobile devices?

- What features could we add to improve the customer mobile experience?

- How do cell phones fit into our customer journeys? Are there gaps that should be addressed?

- How can employees use mobile features to connect with customers?

- What are some ways we can leverage the power of different types of devices that shoppers use?

- What can be done postsale to interact with customers?

CHAPTER 6

Shifting from the Internet of Things to the Commerce of Things: Enter Headless Commerce

Behind-the-scenes changes can provide more flexibility for shoppers and improve their experience.

What happens when you visit a travel platform and make a purchase? While there are aspects that are visible to customers, there is also a flurry of behind-the-scenes interactions taking place. We tend to call what you see the front end. The back end refers to what is invisible to the user, or the server side.

Suppose you want to buy a plane ticket. You head to a travel platform and search for the flight you want. Traditionally, this part of the process involves the back end of the site. You'll interact with the ticketing system to choose your flight and dates. When you are ready to purchase the ticket, the site passes you to a payment system. You don't view these shifts on the front end. However, they are happening in the back.

When you put the name and address on your credit card into the payment system, a number of things can happen. If the information you enter in the payment step matches what was inserted during the ticketing part, you'll continue to move on. You could run into trouble, though, if your name or other details don't match what was entered into the ticketing system. This means if you are purchasing a ticket for yourself, you likely won't have any issues. Problems can arise when you are buying a ticket for someone else. Many of these systems are set up in a way that doesn't allow you to change currency, location, address, or the name on a card. If you're trying to buy a ticket for an older parent or a relative, you might get stuck at this point.

Obstacles like this on the back end of a website can quickly lead to customer complaints and unhappy experiences, which are exactly what we do not want. Fortunately, there are new ways to rework old systems, and they often lead to a better outcome for the customer. Furthermore, they provide more opportunities to touch customers at different points. They make it easier for consumers to complete transactions on their own time and in their own way.

> **There are new ways to rework old systems, and they often lead to a better outcome for the customer.**

Headless commerce, also known as composable commerce, is one way to be more agile in the online space and readily meet customer preferences. We'll talk about what the term means and look at ways to start thinking about how it could be used. With a little dreaming, a lot can be done to provide more flexibility during the shopping experience. Most of it takes place behind the scenes.

Understanding Headless Commerce

Traditionally, e-commerce systems have a front end and a back end. As mentioned, the front end is what the user sees. The back end isn't visible. Typically, these platforms have not allowed for a clear separation between the two. They are set up with predetermined experiences in mind for the consumer. If the customer doesn't follow the exact layout and has different preferences, there is little room for personalization. This causes limitations for what the shopper can do, as we saw in our example of the airline ticket.

Today, these arrangements are changing. It's becoming more and more possible to separate the front end and the back end. If you want to design an e-commerce platform now, you have many more choices available. By making it headless, which refers to separating the front end and the back end, you can create a back end that will work with any kind of front end (or head) that you want. For instance, the back end could support what you see on a website, at a kiosk, in an app, or through a voice assistant device or even a watch. You might even make the front end be a fridge or a car.

These changes mean that businesses can be nimble and integrate into any system. This allows customers to make purchases on their own terms. Through headless commerce, companies can interact with customers in person and via screens. More importantly, they can even engage with customers when there is no screen in sight, as might be the case with equipment such as a voice assistance device.

Leveraging the Capabilities

As technologies continue to roll out, using headless commerce makes it easy to add new pieces into the system, without having to start from scratch. Say an auto has a screen in the front that allows the driver and passengers to interact with the vehicle. Perhaps the device attached to the screen could track miles. It could reach out to the customer when it's time to schedule maintenance, rotate tires, or change the oil. What if the driver or passenger could line up a servicing time for the vehicle or purchase a new set of windshield wipers right from their seat via the screen?

If a fridge comes equipped with a device that can track and order groceries, a customer might ask it to monitor the supply of milk and eggs. What if, when these ran low, the device alerted the consumer? Or perhaps it could even direct an order to the grocery store. Those items could be purchased and delivered to the customer. All these ways can help simplify the lives of consumers and make their transactions come together smoothly.

Incorporating headless commerce into a business plan adds another layer of data that can be tapped. Organizations can look at how customers are communicating and where they are making purchases. Are they engaging in voice interactions? Are they frequently on their phones? Do most of their transactions take place on their desktop?

Headless commerce enables the back end to track what consumers buy. Once this data is captured, it can be used to reach out to customers in a personalized way. What if you realized that a customer tends to buy the same pair of pants, making a purchase every few months? Could a notice be sent their way when a new color of their favorite pants comes out? What about special deals and promotions related to clothing items that would go well with the pants? Managing the data and using it to drive more experiences requires a bit of work, but the payoff can come in the form of higher profits.

Understanding the Commitment

While going headless certainly provides advantages, it's important to recognize that making this transition is not a small effort. It requires a strong dedication and often the budget backing to make it happen. It does create seamless experiences and can improve efficiencies. However, it can also take long hours to maintain. Organizations often don't venture into this design on their own. Instead, the company might bring in a third-party platform to help achieve its objectives related to headless commerce.

For smaller companies with limited resources, the switch might seem too daunting and costly to make business sense. In addition, some organizations aren't ready to look at advanced personalized experiences and customized omnichannel experiences. In these cases, it may not be the right time to implement headless commerce. Larger companies with time and budget constraints, or a vision that doesn't align with personalization and flexibility, might not venture into this space.

When considering the return on investment, keep in mind that bringing in headless commerce could create meaningful and improved customer experiences. Some of the changes that occur when implementing this structure could lead to the opening of more markets. Different types of customers might start reaching out to your company, especially if you provide seamless movement from one channel to the other. The end result may be a larger customer base and more sales … all because you made it easy for customers to interact and move through the buying process in a flexible, personal way.

USE CASE: AMAZON KNOWS HEADLESS

While not all organizations are jumping into headless commerce, some have already made leaps and bounds into the space. Amazon is a standout example of a commerce solution that is designed to accommodate different needs. By separating its back end architecture from its front end, the company is able to provide a seamless experience for consumers across different devices. Shoppers can engage on their desktops, on their phones, or even through Alexa, the voice assistant. A customer might check on an order status through their laptop or simply ask Alexa for an update. These features have helped Amazon become a go-to source for many consumers. These shoppers might not understand the back end features that provide their seamless experiences, but they know how it makes them feel, and that feeling is great.

The entrance of headless commerce may not be visible to consumers on the outside, but they'll notice that they can move effortlessly from one system to another. These shoppers will appreciate the seamless experience they gain. Before you "go headless," consider how it could serve your organization. Carry out a cost-benefit analysis. Then think beyond the current customer base: What other segments might be attracted to more flexibility? How will providing seamless experiences impact the brand and its reputation? What are some ways to implement headless commerce? What benefits can we provide for our current and future customers? You'll quickly find that delving into headless is an exciting phase and one that can result in significant benefits for the organization and consumers alike.

PUTTING IT INTO ACTION

To start thinking about ways headless commerce could be leveraged for your company, spend some time going through the following questions:

- How is our back end and our front end set up?

- Do the current systems cause problems for customers? Which issues arise most?

- When do customers start having trouble interacting with us?

- What can we do about interactions to make seamless experiences for consumers?

- What are the benefits we could gain from implementing headless commerce?

- Which costs and commitments should we consider that are related to headless commerce?

- How could we manage data coming in? What insights could that give us, and how could we use that to move forward?

- Are there other consumer segments we could be tapping by improving the shopping experience with headless commerce?

CHAPTER 7

Let's Talk about Brick and Mortar: Creating Transactions from Experiences

The biggest misconception about stores involves viewing the location as a place solely designed to present inventory and watch customers buy it.

Want people to come into your store? Sometimes it's not that hard. Every day, an average of three hundred thousand people walk along the Avenue des Champs-Élysées in France's capital city of Paris. They take in the monuments that adorn the boulevard, enjoy one of its many festivals that take place throughout the year, or slip into a high-end store while on a luxury shopping spree. The street section runs a little over a mile in length and is frequently called the world's most beautiful avenue.[42] Noted as a must-see for tourists, shoppers often combine a bit of vacation with commerce and appreciate the experiences that come with it.

Many of these shoppers line up to get inside the Louis Vuitton Maison. This retailer has an impressive stance on the corner of the Champs-Élysées and George V Avenue. As the largest location in the world for the brand, the store rises five floors and rests in an art deco building complete with a period dome. The structure is listed as a historical monument and dates from 1914.[43] The store features everything from luggage to purses, attire, jewelry, and leather goods, among other items. While there, shoppers sip champagne while being guided through the selection process by white-gloved store associates. Browsers can glide up impressive open spiral staircases and absorb the history of the brand, which is displayed inside the building. At every step, they'll take in the impressive architecture and design.[44]

The majority of the reviews left by visitors on Yelp about the Louis Vuitton Maison Champs-Élysées gush about the "experience" they had at the store.[45] When I went there recently to purchase a handbag for my wife, I smiled at the orchestration of it all. After waiting in line outside of the store, I was invited into the exclusive, pampered interior. Immediately a store associate showed up with the staple glass of champagne. They led me through a choice of purses, taking out each option with gloved hands and asking me for input to find an ideal fit for the occasion. I am certain the price I paid for the handbag I walked out of the store with was much higher than the cost of manufacturing it. Then again, part of the charge covered the experience of customized treatment and the doting environment surrounding the purchase.

While visitors flock to Louis Vuitton Maison Champs-Élysées year after year, and many find deals that are available only in the store, not every retailer can re-create such a momentous draw for customers. Or can they? I would argue that if we take a step back and evaluate the purpose of a store with new eyes, we can start to envision different approaches. Think of it as hitting the "reset" button on stores. Do they have a place in today's commerce world? If so, what should that place look like? How should it interact with customers? What should it sell, and what can it offer? What experiences can it create?

All these questions help us dive into the debate over brick-and-mortar stores in commerce, a hot topic in recent years. As we saw in earlier pages, a plethora of retailers have closed shops that may never come back. While the headlines tend to tout the doom-and-gloom aspects of closing physical locations, there are genuine opportunities to be found in this new postpandemic world of shopping. Certainly, we've seen that online and mobile play a role in the customer experience. The store does too.

We would be remiss to focus merely on why stores should—or shouldn't—exist. We also lose sight of the mark when we excruciatingly review the number of transactions made in stores. The best way to view brick-and-mortar places is to consider the purpose they serve. There are so many ideas and possibilities once we shift away from the traditional concept of shelves filled with stock and customers lined up at the cashier to check out. We can start to imagine the

There are genuine opportunities to be found in this new post-pandemic world of shopping.

experiences we want to create and how that will benefit our customers. We will think about how stores will impact our sales—wherever those transactions take place. We'll also be on the right track when we take an optimistic stance on change and blur the lines separating online sales, mobile sales, and store sales.

Experiences Reign Supreme

In terms of the five senses, what can stores provide that online interaction and mobile engagement lack? First, and perhaps most important, is the chance to stimulate the senses of touch, smell, and taste. In addition, while customers can see displays on computer and mobile screens, they can gather an in-depth view of the brand when they step inside the store. Furthermore, they can listen to music, the sounds of other customers, and the overall vibe of a place when they are there in person.

The Louis Vuitton Maison Champs-Élysées example provides great insight into the power of experiences. However, when sharing my story about shopping there for a gift, I always like to point out that zooming in too much on this model can create a false pretense. This is because it's easy to look at high-end retail and assume that experiences only apply to luxury brands. They can set a high price and margin, the thought goes, to cover the underlying expenses needed to create the finesse that those with purchasing power want.

To clarify this misconception, let's look at how experiences could be used within industries that operate at different price points. We'll consider selling groceries, kitchen supplies, and statement watches. Use the following scenarios to help trigger your imagination and start to dream about the events that could be held in your own stores.

IN-STORE DEMONSTRATIONS

Have you ever walked through the aisle at the grocery store that contains spices? Whenever I take the time to peruse this section, I end up feeling lost. Certainly, I can identify basic spices that I am familiar with seeing in the dishes I eat. However, in a well-stocked and diversified grocery store, there are always several spices that I wouldn't immediately buy. Perhaps they would be the perfect fit for my tastes, but I don't readily know. I'm unaware of what they are like and how they are used in the cooking process.

Close your eyes with me for a moment and imagine we are taking a stroll through a grocery store that understands the customer experience and how it impacts their bottom line. (I'll push the cart; you carry the list!) We get to the spice aisle and grab onion salt, paprika, and pepper, which are on the list. Into the cart they go. As we get to the end of the aisle and turn to head toward our next destination, we come across a store associate, a small stove setup, and a spice display. There is a counter to the side of the stove where the associate is placing samples of a dish created with the same spice that is on display. The employee chef invites us to try a bite. Then they hand us a recipe card that lists the ingredients. While we taste, we learn about where the spice comes from and how it is commonly used. The food is delicious, we agree. Better yet, the instructions to prepare it at home look easy. Suddenly the new spice goes into the cart. You add a few items to the grocery list based on the recipe card, and that dish—complete with the newly purchased spice—becomes our dinner in the evening.

Would we have ever purchased the new spice if it were merely placed amid the rows of other herbs? Probably not, assuming we were unfamiliar with it and not looking to try something new. What about if it had only been on a display case in a prominent position of the grocery store, sans an associate to explain its use and provide a demonstration on cooking with it? My guess is no. And if an employee chef was making it, but no other cooking instructions were given, would we have been lured in? Probably not. After all, we may have liked it but not felt sure about how to cook with it at home.

It was the entire experience—the preparation details, the store associate to present it and provide background, the sample, and the easy-to-reach display case—that made a difference. And even if that spice were a bit more expensive than the others on the shelf, it probably wouldn't completely alter our grocery budget.

My point here is that the experience was incorporated into a low-priced store and a transaction resulted. Supermarkets are known for their tight margins. Yet through this example, we can see how a few tweaks can elevate the customer's shopping trip. The display provided new and educational insight into spices, a fun food item to try, and a recipe that offers a chance to develop cooking skills for the dish.

Grocery stores can venture beyond the spice demonstration too. They could offer a cooking class that features some of the ingredients sold at the store. Guests could sign up to learn about Italian cooking methods or to see how sushi is rolled. The event could also consist of a workshop that features the history of Middle Eastern tea. Keep in mind that attendees may be willing to pay for the experience, which could help cover costs or even produce a profit.

These examples show that a grocery store can help customers discover new products and information. It can also provide an engaging experience for shoppers. The actual shopping and replenishing of supplies could take place behind the scenes. Customers might order from home on their devices. If a system monitors their consumption, items that are running low (think milk, bread, butter, etc.) could automatically be purchased and delivered.

SAMPLINGS AND CLASSES

Suppose you are on vacation, spending some quiet days at a retreat just outside a small village nestled in the mountains. The area is known for its appeal to guests who want to get away from the city noise and have some peace and quiet, with few people around. You stroll down the main street of the village one afternoon after finishing a lingering lunch at a mom-and-pop restaurant. Along the way, you notice a store that sells kitchen gear. Its windows boast utensils, cooking mitts, small appliances like blenders and coffee pots, and glassware.

You are about to pass by, as you're not shopping for kitchen supplies that day, when you notice a sign indicating there will be a wine tasting that very evening. You'll be around and don't have any plans. You read on past the top headlines and learn the event will run from 7:00 p.m. to 9:00 p.m. It will feature local wines, with a presenter to explain their backgrounds and features, along with pairing samples. There is a cost of twenty dollars to attend the event, which includes the beverages and snacks served. Visitors are invited to sign up inside the shop or online. There is a limited number of spots available, the sign reads.

You opt for online registration. You pull out your phone and open the website indicated on the poster inside the store window. The information online matches what you see in the window. You enter your first and last name and your phone number and then continue with the payment process. You learn there are twenty spots available, and you take the twentieth position. *What good luck! I'm in!* you think. You tap on the final button to complete the transaction.

At seven that night, you show up at the store and join the other guests who are already inside. True to form, there are wine samples available, along with bits of prosciutto, melon, red meats, and toasted bread topped with tomato slices. A guide walks you through the history of each wine—there are four in all. You learn about how they are produced locally at family vineyards. The pairings are delightful, and you talk to a few of the other guests. They are equally intrigued with the presentation and display.

As the clock strikes 8:30 p.m., the sampling session draws to a close and guests are invited to browse the store on their way out. By this point, you're feeling satisfied with the tastings and appreciative of the wine this region produces. You've also had an enjoyable evening meeting other visitors who are on R & R getaways of their own. On your way out, you notice a few coffee mugs that would make great gifts. You also see a set of kitchen towels and remember that your spatula at home recently broke. You find the section featuring kitchen utensils, grab one, and take it. Together with the towels and mugs, you carry the merchandise to the self-checkout station, which is located near the store exit.

Several days later, you leave the area and head home. The kitchen shop continues to text you occasional promotions. It also asks you for feedback on your visit. In the months to come, you sometimes buy a few more gifts through the place, as items can be sent directly to your home and, if you spend a certain amount, the shipping charges are waived. Since you became familiar with the store during your initial visit to the wine tasting event, you understand the type of products the place offers both in store and online. They mesh with your needs, and if you get back to that village on your next vacation, you'll probably check online to see what's going on at the store. If there are upcoming events at the kitchen retailer that coincide with your stay and interests, you'll likely attend.

A DEMONSTRATION ON WATCHES

Who needs a watch anymore? If we're talking about the original iteration of the device, i.e., the time-tracking tool, we would most likely agree that the lone function of timekeeping is no longer a huge draw. All we need to do is pull out our ever-present phones or glance at the myriad of digital clocks displayed in our autos, grocery stores, retail stores, offices, clinics, and educational facilities. In other words, it's generally easy to find the time when we have so much access to these tools.

Still, watches continue to sell. In recent years, there has been a shift in the industry away from the standard time-tracker feature to other areas. A watch can monitor your steps, allow you to make phone calls, provide games when you need a distraction, and keep track of your health, among other functions. Besides the heavy-featured watches, there is the status symbol appeal. A high-end watch reflects a certain social status and income level, and consumers who want a luxury look on their wrist are attracted to these shiny objects that happen to tell time and might carry out other tasks.

If I decide I want a watch, I can go online and find a multitude of options. I might be able to buy one within a matter of minutes. I can choose to have it delivered to my home within a couple of days. When it arrives, I can put it on right away.

There's just one problem. Will I know how to use it? What if this time tracker is one of the latest smartwatch versions? If I haven't done any research, I might not know all the smartwatch's features. I may be unaware of how it can best be used in my daily life.

Enter the watch store. This type of retailer could have an important purpose. Rather than trying to sell as many watches as possible, the shop could take a different approach. It could help consumers understand how their watches function and how to maintain them.

Continuing with this thought, perhaps a retailer sells smartwatches, pilot watches, sport watches, and diving watches, among others. Its sales staff is well informed on how each device works. When a shopper comes in and asks a question, the sales associate has a knowledgeable answer.

There's no sense of rush for consumers. Instead, sales associates are ready to take the time to compare watches. These representatives are quick to ask questions too. They'll want to know why the consumer wants a watch. They'll inquire about the consumer's interests and hobbies. They'll also ask about price points.

The answers to these questions will guide the merchandise the company representative recommends. As they show options, they'll also discuss the features each one has. They will demonstrate how it works and best practices for keeping it clean. They'll go over battery details and talk about how to replace or recharge the watch's energy source.

Influencers can play a role here too. They might come to the store and carry out a demonstration of their favorite watches. Or they could create posts on social media that show how their new watch functions. They might even share a list of up-and-coming watch designs.

The store could hold events such as a presentation from an Olympic athlete. Perhaps the discussion centers on a certain sport, such as swimming. It could include recommendations for watches that work well in the water. Customers could buy tickets for the event. They might be promised an autograph or picture if the presenter is well known in their industry.

COSTS VERSUS SALES

These examples of the grocery store, kitchen store, and watch shop share some similarities. They all provide an experience for customers, albeit in different forms and at various price points and margins. You may have also noticed another common feature: budget-wise, there are costs to putting them on. The grocery store will need an extra worker to present to customers, along with the ingredients and eating utensils necessary for the samples. The wine event includes costs for a presenter; the wine consumed; disposable plates, cups, and napkins; and the accompanying pairings. The watch shop will have to compensate models and arrange attire and accessories or hire an agency to help oversee these aspects. Diving into expenses further, we might say there are potential utility costs if the spaces are open and being used outside of normal business hours.

Now let's turn the table and look at the upside. The grocery store could sell the spice being featured for a higher margin. This could cover the demonstration costs and still bring in more revenue. The wine event charged a fee, which shows how customers may be willing to pay for certain services. This is especially true if the event involves an experience that they'll find gratifying or that will take them through a discovery process. The kitchen supplies store could make sure the cost for the wine tasting not only serves to pay for the related expenses but also brings in a profit. And the watch shop? It can charge for sport-related events, especially if they feature an accomplished and well-recognized athlete.

Embracing Change

So often it's easy to fall into repetition when it comes to online and store setups. It's natural to try to replicate what a customer sees online with what they view when they walk into the store. Since we've observed that different senses tune in to the different modes of shopping, it is logical to look at stores in a new way, and one that meets customers' in-person needs. Online may be great for customers who know what they want and just need a few minutes to quickly check out. In-store shopping can be ideal for experiences, relationship- and brand-building events, and the chance to touch and feel merchandise.

Too often I see an emphasis placed on the division of transactions. What if 90 percent of the purchases are made online, and 10 percent take place in the store? Is that enough to justify a store closing? Perhaps yes ... and perhaps no. I like to advocate for a holistic approach rather than a competition between online and in-store transactions.

If the store is used to create an experience, the shoppers who attend events in the store walls might decide they love the brand—and buy from the company in person or online. Take Apple, for instance. The company has greater sales figures per square foot than any other retail location.[46] Its sales associates introduce products to customers, upsell, and originate many purchases in store. Shoppers who don't buy anything at the store might go to their homes and purchase what they saw later.

> In-store shopping can be ideal for experiences, relationship- and brand-building events, and the chance to touch and feel merchandise.

It's typically best to evaluate how online and in-person commerce complement one another. A solid framework will incorporate both aspects. It will make it almost effortless for a customer to find their way and complete a transaction through a website or store location.

While there is a place and purpose for stores, we also need to recognize that making a case for stores and experiences shouldn't end in an organization opening dozens of shops with little forethought. Nor does it imply that the existing stores are essential and should be left open. The best approach will involve a review of the store locations and purpose to decide their value. Are they in the right spot? Does the data show that customers of the brand currently live far from existing brick-and-mortar locations and would prefer a shop near them? Are there currently three stores in a certain area that only needs one local presence?

All these are good questions. However, I tend to encourage clients to look at them from a different perspective. By this I mean we need to take the focus away from the transaction. We must center it on the experience. Answer the questions by thinking about how in-store events can be created. Reflect on how the store can be a useful place.

We might find, through this exercise, that the store has a much bigger purpose than we realized. In all cases, we'll want to view stores in the context of the entire customer experience. We won't want the word *transaction* to be at the center of our decisions.

One final note on opening and closing stores. With today's tools, we don't necessarily have to hire and let go in traditional ways. If a store opens in a new location, it may be worthwhile to see if remote workers live in the area and want to come into a physical spot. When closing a brick-and-mortar space, employees might be able to shift to online work. Perhaps they help with customer support and stay in their homes. This can allow them to avoid looking for a job in a different company. They can stay with their current employer and not have to move to another location where a physical store is still open.

Building Communities

Perhaps the best part of having a great store, and the most exciting aspect of keeping doors open, centers on the concept that consumers prefer to shop with brands they like. They're looking for retailers that share their values and make them feel good when they step through their doors. In fact, research shows that when consumers appreciate a company's cultural richness, they prefer to shop in store instead of online.[47] Consumers are attracted to the traditions, history, and ideals that brands portray.

Just think of customers who shop at grocery retailers like Whole Foods over other establishments such as Walmart. While these each have a place in the market, Whole Foods caters to shoppers willing to spend a bit more on products. These consumers are potentially looking for organic options and value nutrition. They also want an environment that is clean and friendly. The experience comes into play in several forms. Shoppers can stop at the café at Whole Foods and have a meal during their outing. They can browse vitamin displays and learn about what they offer for their health. They can leave feeling like they purchased foods that are good for their well-being.

Starbucks is another example of an establishment that provides an experience and draws a crowd of consumers who love to order their beverages—or customize their own. Influencers touting the latest drinks available at Starbucks abound on social media. These online stars also regularly list their top beverage picks. Young people visit and post their selections to their Instagram accounts. Their friends will view these stories and potentially order the same item next time they go. You'll find many Starbucks nested into bookstores, which appeals to those who want to study, read, shop, unwind, eat, or drink … or, most likely, a mix of several of those options.

Amazon, which already has an expansive web footprint, is making a carbon one too. The retailer is opening new physical locations. These places will showcase high tech solutions, including an app that integrates virtual and real-life experiences. Customers can make clothing selections in the app. These pieces are automatically taken to the dressing room in the store. When shoppers arrive, they can head to the fitting room, where their options await.[48]

USE CASE: WAYFAIR ADDS STORES TO ITS MIX

Remember the online furniture store with options galore for mobile users? In December 2021, Wayfair announced they would open a new kind of store in three locations. The company also presented a plan to develop additional stores during the coming years. If you walk into one of these locations, don't expect them to be your typical furniture store. "We are introducing a new kind of omnichannel shopping experience powered by the Wayfair platform, inviting our customers to engage with the brands they know and love in an innovative format that blends the best of in-store and online shopping," stated Karen McKibbin, head of physical retail at Wayfair.

Prior to the plan to open more stores, the company had operated a variety of pop-up shops and a smaller retail store. With the next steps, customers will be able to start their experience online. They can move into a physical location if they choose. Then they can complete the transaction in whatever way feels best to them.

It's time to rethink how we use stores and the experiences they create. Just as mobile is important to include in the customer path, retail shops hold their place. The key is to explore the why behind the doors and walls of a place. If it's a good fit, the store can serve as a place where customers learn, discover, get entertained, and walk away with a better connection to the brand.

PUTTING IT INTO ACTION

To properly evaluate the way we treat stores and the purpose they serve, spend some time reflecting on the following questions:

○ What store locations do we currently have? What are they like?

○ How do our brick-and-mortar locations compare to our online site? How are they the same? What is different?

○ What sort of experiences could we create in our stores that would add value for customers?

○ What would the costs and benefits of these experiences be?

○ Do we need our current locations? Are there changes to be made?

○ How can we best utilize employees during transitions, such as store closings?

○ How are our mobile, online, and in-store experiences related? How do they fit into the customer journey?

○ What are our company values, and how do our stores reflect that in a way to bond with customers?

CHAPTER 8

Virtual Space and Commerce:
A How-to Guide

If you haven't yet put on headgear and dipped
into a 3D experience, or spent a few hours
playing video games, now's the time.

Did you raise your eyebrows when Facebook rebranded to Meta? If you questioned the move, you weren't alone. If you weren't surprised, then you're like me. Though some of its employees balked and even stepped away at the time, Facebook-turned-Meta forged on with its evolvement. Founder and chief executive of the company Mark Zuckerberg explained that the new brand would introduce people to shared virtual worlds and experiences across various platforms.[49]

The company then made a series of moves to leave no doubt that it was all in with this new vision. It hired thousands of employees to fill newly minted positions at labs that develop hardware and software for the metaverse. It has picked up engineers from rival companies like Microsoft and Apple.[50] Some employees are deeply entrenched in jobs that call for them to reimagine how concerts and conferences could be experiences. The company plans to take on ten thousand workers for its metaverse construction in the European Union during the coming years.[51]

Curiously, this isn't the first time Facebook has overhauled itself. In 2012, the company underwent a transformation to focus on mobile users. This move led to a long series of successes and expansion. At the time of the change, cell phones were widely circulated and used. Thus, the move to make Facebook mobile-friendly seemed like a natural next step.

Now, as Facebook looks to build and expand the metaverse, the landscape isn't the same. Unlike the hugely popular mobile devices that existed in 2012 and spurred change at the company, today's metaverse isn't exactly fully constructed and thriving. This lack of foundation on which to build and branch out has some worried. They are wondering what the future will bring. Will the metaverse take off as Zuckerberg is betting? Or will it be a trend that never quite develops into its full potential?

These are serious considerations to keep in mind as we face the coming years. Though I understand the concerns surrounding a technology that comes across as fuzzy in mainstream channels, I also regularly advise others that Meta's moves are a big deal. The company is a massive force. It has invested billions of dollars to carry out its plans for the metaverse. It has the power—and desire—to change markets.

When it comes to virtual spaces, we can better understand how they work and what they mean for commerce by breaking down the topic a bit. Let's spend some time looking at what the metaverse is. We'll also consider some examples that show how virtual experiences are popping up and have potential for commerce. We'll close with a fun thought, which underscores the value of play for stimulating creativity in the business realm.

Emerging into the Fascinating World of the Metaverse

It's a video game … no, it's a buzzword … or better yet, the next internet phase. All these terms have been used to describe the metaverse. Many people have still come up stumped when trying to decipher its full meaning. Conversations about the virtual space almost eerily mirror many discussions about the internet in the 1970s and 1980s. During those decades, there was little knowledge about how it operated and what exactly it held for the future. As we reflect on the past decades, we can see that some of the internet's expectations held true; others did not.[52] Its development can help us gain perspective on the metaverse. Since it is still evolving, some of the theories about it may turn into a reality. Others, however, won't.

For the sake of our topic on the metaverse and commerce, let's move away from some of the fancy phrases we hear about it. Instead, let's focus on what we can currently see. This consists of a headgear that provides you with a 3D experience. You can interact in this space. The setup has been around in some industries, such as gaming, for years. Recently it has been shifting to become mainstream. You might spot the chunky headgear when walking through the mall, strolling through a fair, or attending a conference. This equipment is out and about, though not every home on the block has their own headgear ... yet.

Consumers Love Virtual Spaces

If you've ever visited a shopping center and spotted booths filled with individuals sporting virtual reality headgear, you may have heard a plethora of sounds. These could range from shouts of excitement to screams of terror. Is everybody okay? By and large, these consumers are doing just fine ... to a certain degree. They are typically in a safe, clean environment and aren't facing any grave physical dangers. That said, their minds think they are going through an actual fall or fight. The technology is so advanced that the movements seem real. When visitors face a fall in the virtual space, their brains perceive that they are tumbling down in real life. The experience ends, the headgear comes off, and the customer walks away, unscathed. (Though I'm guessing adrenaline pulses through their veins for a good bit after it's over!)

This is the power—and the allure, in many cases—of virtual reality. If you head to Las Vegas, you can choose from a long list of virtual experiences. Want to travel to an immersive universe where you can experience adventures with your friends? Go to Virtual Reality Powered by Zero Latency at the MGM Grand. While there, you can fight off killer robots and skirt along twisty pathways. How about a roller coaster with a virtual reality headset you can wear so it feels like you are flying over a futuristic Las Vegas? Head to the New York New York Hotel and ask for the Big Apple Coaster. Or are you feeling like an escape room equipped with virtual reality features? Sign up for the Virtual Room at Madame Tussauds.[53] The list goes on. Every year, tourists flock to these attractions. They thrill in the experience and often leave with a desire for more.

So how do these adventure virtual reality trends translate to customer experiences in commerce? I think they show an openness and willingness to participate in other spaces. Consumers are also growing in their awareness of and level of comfort with the technology. Some of the features that were previously only known to heavy gamers are now going mainstream.

The technology has advanced too. Headsets have gotten smaller during the last years. They are less clumsy to put on and take off. They can be comfortably worn for a longer period. These attributes make it easier for consumers to latch on to the concept and accept it.

> **Some of the features that were previously only known to heavy gamers are now going mainstream.**

Let's also add that customers are not only buying tickets for these augmented rides. They're also putting on other devices laced with virtual technology. Ray-Ban, a prescription eyewear brand, offers customers smart eyeglasses and sunglasses. Called Stories Smart Glasses, this eyewear comes equipped with camera, audio, and meta technology. Customers can put on their shades. Then they take photos and videos, listen to music and calls, and even share content on their social media pages.[54] There are other glasses on the market and in development. They are all designed to add layers of technology that are easy for consumers to access and use.

Taking this a step further, we can imagine how virtual reality and headgear can play a larger role in commerce. Think about the big advantage these spaces hold. They allow users to experience a time and place that is outside of where they are physically located. This capability can be used in a variety of industries that stretch beyond games and entertainment.

If you're a travel company and are marketing high-end destinations or big-ticket experiences to your customer base, why not send a potential client a headset? They can use it to get a peek of what their next adventure could look like. Say they are interested in a trip to Mount Everest but aren't quite ready to pay the $15,000 fare. They receive a headset, put it on, and immediately feel immersed in the thrill of the climb. They're surrounded by majestic peaks and a white landscape. Or what if you are selling a stay at an all-inclusive resort in the island of Saint Martin in the Caribbean? You could ship a headset to a prospective couple who is considering spending $10,000 on the trip. They can experience the sand at their feet, the waves against the shore, and the quiet escape that could be theirs. All they have to do is sign up for the vacation.

On a similar vibe, if you're a car manufacturer and are in the process of rolling out a new model, you could leave headsets with dealers for customers to check out. They might be interested in coming in to "see" and "feel" what the inside of this up-and-coming auto is like. After experiencing it through virtual reality, they could preorder the vehicle. Then they could tell their friends their new car is on the way.

While the addition of headsets can enhance a customer's journey, it should be noted that the technology's cost will need to be evaluated. Organizations with a budget that can handle bringing on headsets may find they are a good fit. The same is true for companies that primarily sell high-ticket items. The price point for headgear has dropped in recent years. As it becomes more widely used, the cost of acquisition may continue to come down, making it more accessible.

Haven't Played Video Games in a While? Now's Your Chance!

When is the last time you spent a few hours on Fortnite? What about donning a headset and playing a virtual reality game? Many people would call out these activities as unproductive. They would add that online charades are definitely not something to engage in while at work.

I tend to find that playing games with the right approach can have the opposite effect. While I'm not suggesting we spend our weekends attached to the Xbox in our living room, I do feel it is beneficial to know what's out there in the gaming space. Immerse yourself in a game, and you'll catch glimpses of what the technology can do and how you interact with others in the virtual space. Moreover, consumers are regularly tapping into these games. From building structures in a virtual world to running a farm, users are dedicating portions of their lives to these spaces and are becoming familiar with what they can do on them.

Once we grasp what these games offer, we can start to think about how the features could be used in commerce. It also stimulates creativity in a positive way. When we are so deeply involved in work activities, reports, and meetings, it can be easy to see our imaginations go unused. Tapping into video games allows new ideas to come to life. The play enables concepts to dance around in our heads. This creates the chance to bring innovations to the office and business.

> **Immerse yourself in a game, and you'll catch glimpses of what the technology can do and how you interact with others in the virtual space.**

Some retailers have already moved forward in this realm of virtual commerce and play. The beauty brand Charlotte Tilbury has a virtual space where shoppers can complete a challenge. They are asked to find three golden keys to unlock access to a specific lip color.[55] Consumers can also invite others to join their virtual store experience through the Shop with Friends feature, which is sent via text or email. Once friends join, the group can navigate the virtual environment together or on their own, much like they would in a multiplayer video game. Shoppers can also view the live video feed of their friends on their screen or hide the feed to fully immerse into the virtual environment.[56]

USE CASE: CLIMBING OR FALLING: YOUR PICK

Red Bull, known for its energy drink, offers visitors some intense virtual reality and augmented reality experiences at the Swiss Museum of Transport in Lucerne. If you go, you can experience what Red Bull athlete Jérémie Heitz went through when he summited the most inaccessible peak of the Alps. He climbed the Matterhorn Mountain. Now you can put on a headset and experience that same ascent—virtually. With a harness in place, you'll scale a wall and be rewarded with breathtaking views once you get to the top.[57]

If a plunge is more your style, you can head over to the Red Bull Stratos station in the same museum. The experience is based on Felix Baumgartner's freefall that broke the speed of sound. After launching from New Mexico, Baumgartner ascended to the stratosphere in a helium balloon and then jumped. Now visitors can experience the same sensation he felt through virtual reality. They simply put on headgear and step inside a capsule.[58]

These attractions reflect an insightful view on the part of Red Bull into their customers' interests. Those who consume energy beverages tend to be looking for an extra boost. Based on the concepts of exploring, pushing boundaries, and engaging in powerful events, these experiences provide users something they will long remember ... especially when they reach for their next energy drink.

The first step to understanding the metaverse is to be aware of what's in the marketplace. Try on a headset, jump into a virtual experience, and play video games. As you do, study what exists and how it's being used. Then take a step back and dream about ways these new technologies could be implemented into your brand. Think of how they could create experiences that provide value for customers.

PUTTING IT INTO ACTION

As you decipher the best ways to maximize metaverse capabilities, consider the following questions:

○ Do we currently utilize virtual spaces?

○ What do we know about our customers regarding their use of headsets?

○ Which experiences could we create that would be helpful for customers?

○ What are the costs and benefits related to investing in headsets for customers?

○ Would headsets be a good fit in our stores? Would they work to send home to customers?

○ If we had no limits, what would we suggest doing to make the most of virtual spaces? Are there ways to accomplish some of those ideas with our current resources?

CHAPTER 9

Engagement Everywhere: TikTok, YouTube, and Building Experience-Centered Communities

If you do not have an engaging, video-based way to interact with customers, you are not part of the conversation.

Do you have an extra ten minutes? You can catch a quick tutorial on how to maintain the grill you just purchased. You might watch reviews of the latest grill covers to protect the equipment while it sits outside. Or watch a recipe that you are planning to cook tonight. A few minutes in, and you'll have a strong grasp of what you need to buy to keep your grill in tip-top shape. You'll also know what ingredients to grab at the store before dinner.

Welcome to the world of video and all that it offers. For consumers, it's a way to access information, be entertained, and participate in a community. For companies, it offers a vast assortment of possibilities. You can build your brand, engage with customers, make great use of your ambassadors, and have fun at the same time.

There's so much to engage with in our world, and it goes beyond video experiences. There's a component in video games that provides this interactive, immersive atmosphere we so often crave. Let's break down what video means to companies and their customers today. It's also worth looking at the benefits that can come from reliving our childhood and spending the afternoon playing a video game. (Yes, you read that correctly—I'll explain later in this chapter. Keep reading and get the Nintendo ready!)

Understanding TikTok, YouTube, and More

Read through an online ad or article on a topic that interests you. Then page through a printed brochure. Finish by watching a video on YouTube. Which one provided the most interactive experience? What did you most enjoy? If you answered with the video option, you're joining ranks of consumers who are turning to video to find the information they need and engage with the brands they love.

There's so much more that can be communicated in a video. Think of influencers who feature women's fashion. They can post pictures of themselves wearing the latest J.Crew designs. They may also chat in a podcast about their luxury handbag collection. With both these means of communication, viewers and listeners will likely feel limited in their understanding of the products. There is only so much you can convey on a static web page or through audio.

Compare this to video. Consider the YouTube channel that a women's fashion influencer runs. Suddenly we can see the influencer model those J.Crew designs while she walks through her backyard, cooks up dinner, or goes out to the theater. The same is true when fashion influencers post on TikTok. Perhaps they present a clip on luxury bags. We can observe different angles of top-end purses and get an idea of their size by watching them being carried in a short video.

In addition to YouTube and TikTok, other social media platforms have caught on to video's popularity. Want to share a quick video on Instagram? No problem. Interested in posting a clip on Facebook? Go ahead. Be prepared to watch other platforms surface in the coming years. The demand for videos and the engagement they bring will continue to rise.

Getting into this realm doesn't involve grasping how a video is made and posted. Indeed, those steps are rather minute when we consider the bigger picture. We need to think about who is making the video, why they are creating it, what it is discussing, and how consumers can find it. That's just to start.

Besides making it easy for people to subscribe and be notified when the latest video comes out, we'll want to monitor comment sections. We can also make it easier for consumers to purchase products. YouTube, for instance, has features that allow creators to post links so viewers can locate the products featured. Those shoppers can then purchase the items. Instagram and TikTok have shopping capabilities built into their platforms.

These tasks don't have to be complicated. They also don't need to be cheesy and overly promotional. The first step to creating video consists of developing a framework. This should encompass the purpose of the video channel. Think about how you can fill a need. Then set up a specific channel that aligns with your customers' preferences and behavior.

To see this at work, let's consider running-related gear. If your brand sells fitness shoes and clothing, you might develop a channel on YouTube designed for a running community. You can post videos that feature great workout trails. These clips can discuss future races in your area or take an inside look at an upcoming pair of shoes that hasn't been released yet. The trick is to evaluate your customer experience. Then you can look for ways to improve it and make it more interesting.

For commerce, video helps us achieve our goals when our content is portrayed in a real, relatable way. There's no need to make exaggerated claims or hide flaws. Customers can quickly spot when something is fake. When they see that something is staged, it's a big turnoff. On the other hand, if retailers come across in a human, personal way, the audience notices and typically rewards those brands for their authenticity.

Building the Community

If we are working with ambassadors, these individuals can be a great starting point to build community in the video realm. They may already be active on a platform and have a large audience. If they aren't active, they could be interested in getting started. Those who don't have a large audience might want to develop a niche base for a specific product line.

> **For commerce, video helps us achieve our goals when our content is portrayed in a real, relatable way.**

You can also research your customer base. Check to see if they tend to watch YouTube or TikTok, hang out on Facebook, or scroll through Instagram. There are social media listening tools that will tell you where your brand is being mentioned. You can check what is being said. You can also decipher whether the conversation is positive or negative. These clues will help you decide how you want to use video and where you want to place it.

One of the main ways to create engaging videos lies in shifting away from a transactional mindset. We shouldn't start producing clips if we are only thinking about getting items into the online cart or out the brick-and-mortar door. We'll want to contemplate how videos can be incorporated into our overall customer experience. Should they be used to raise brand awareness? Inform customers of the latest releases? Show how to use our products? Or all of the above? As you look at your overall goals for reaching customers, you can identify the purpose for creating videos.

USE CASE: GILLETTE: ALL YOU NEED TO KNOW ABOUT SHAVING

If you just purchased a Gillette razor or are looking for shaving tips, the brand has ample resources available on YouTube. Here are a few you can watch:

- "Frequently Asked Questions on How to Shave with Gillette"[59]

- "Gillette Fusion—Do These Things Really Work?"[60]

- "How to Shave—Shaving Tips for Men"[61]

- "How Gillette Razor Blades Are Made"[62]

- "Gillette Razors—Types of Gillette Cartridges / Differences in Razors"[63]

Look at these videos or browse for others, and you'll observe that Gillette creates some of their own videos. Others are developed by influencers and brand ambassadors. Together this mix provides a down-to-earth, accessible resource pool for those who want to know about shaving. Customers can decide which razor is right for their needs, how to use what they buy, and all sorts of other advice to improve their overall experience.

Communities might start slowly and grow over time. They could also get a jump-start, especially if their commencement coincides with other initiatives, like a big event or a promotional piece. To make them successful over time, they must not be ignored. We will want to moderate the community, look for feedback, and adjust our approach as needed. We shouldn't expect the group to grow on its own. It will need ongoing attention. This commitment and engagement could come from our own staff members, our ambassadors, or both.

The Benefits of Community

It might seem like a lot of work is needed to create a following. I typically agree that there is effort involved. I would be quick to point out, however, that the benefits can be vast and long lasting. More engagement can lead to more repeated customers, which will lower the cost of customer acquisition. Word of mouth will take care of the marketing components. In fact, video is the cheapest way to bring new customers to your brand. It also is among the best ways to increase your number of repeat customers.

There's so much valuable information that you can gather from videos. Post one, and you'll quickly see how many viewers have watched it. As you create more, you can evaluate which ones were most popular. (There could be some surprises—a video you thought was rather dull could take off!) You might add on hashtags and check to see what's trending. Of course, there is the opportunity to go viral, and videos that raise a ruckus often teach us what consumers want.

There's also the dual benefit of working with ambassadors. Say your company sells skincare products. A dermatologist might be interested in sharing information about your new sunscreen, especially if its ingredients meet her approval or if she recommends it to patients who come into her office. Through educational videos, the doctor can build her own brand, raise awareness about sun protection, and promote yours—all at once.

The shelf life of videos makes them accessible, attractive, or both. Instagram allows you to post a video that disappears after twenty-four hours. The shortly lived clip can stir feelings of urgency and scarcity among consumers. This might make them eager and willing to click on the button to see what's available—before it's gone.

If you post videos on a platform that doesn't cause them to evaporate after a day, you'll have the benefit of building a library for customers. These videos can be grouped and sorted. They could also be shared with others at appropriate times. For instance, as the holidays approach, an instructional reel on how to wrap a gift could be passed around and sent to customers. It might not matter if the video was created in 2012. If the content still feels fresh and relevant, the evergreen video can be used year after year.

USE CASE: GOPRO: ADVENTURES IN TIKTOK AND BEYOND

Interested in what it feels like to own a GoPro? This company specializes in adventure cameras. Head to TikTok and look for the brand. Along with its millions of fans, you can watch the company showcase its cameras.[64] Made to capture the world, this versatile device can record sports in action or any other immersive or exciting endeavor you take it on. Footage from the camera can be uploaded to the cloud, and you can get a firsthand look at how the camera is used by watching videos of the equipment in action. From cute to creepy, thrilling to heart jumping, TikTok consistently releases videos to engage with potential and current customers. And that's not counting the other influencers, ambassadors, and users who are sharing their GoPro adventures with the world as well.

Take Time to Play

Now that we've looked at creating video, let's take our discussion up one notch and consider the in-game experience. Many brands have started testing out this new realm. They've developed new ways to build their brand and grow their business. The idea is to give customers a fully immersive experience that is impactful or meaningful for them.

While this may sound like an interesting concept, you might be wondering where to start. I often advise individuals who ask me this same question to play an immersive video game. I think a key lesson can be learned from spending a few hours on a weekend at the Xbox. It lies in the chance to step away from the to-do list, to think outside of the processes that tend to guide our days, and to delve into a world of creativity.

This can get our own imaginative juices flowing. It also helps us think about new ways to carry out business. For instance, how are purchases made when playing Fortnite? Why do customers want to buy equipment and gear for the game? What benefits do they receive from the playtime? Are there social aspects that are being incorporated? How does the customer experience translate to a video screen? What feelings and atmosphere are tied to it (i.e., the allure or draw for customers)? Perhaps you'll agree with me that much can be observed, from a commerce standpoint, by playing a video game!

Of course, Fortnite is only one example of many options on today's market. There are others that allow you to step into an immersive experience and create in new ways. Think of Minecraft, in which you can play to survive, battle mobs, build your own shelters and structures, and explore the landscape.[65] The game has become a popular educational tool. Teachers encourage their students to create their own spaces and build their design skills. There are collaborative elements, as kids can play the game alongside others and work together to make their communities.

There's also Stardew Valley. This game begins with you inheriting your grandfather's old farm plot. You start your new life there with a few rustic tools and coins. You then push against the new Joja Corporation, which is making the old way of life slowly disappear. You aim to restore the valley and make it shine again.[66]

Games like Stardew Valley are instructional to an extent. They provide users with an in-depth learning experience. On the farm, you need to buy certain pieces of equipment. You must know how many supplies to get. And that's just the beginning. Over time, you can find a partner, start a family, interact with other characters, uplevel and specialize your skills, and even learn how to cook!

I spoke with a twelve-year-old boy who played Stardew Valley. He explained to me how he was setting up his farm. He outlined the types of vehicles he purchased, including a certain kind of truck that could carry the goods he needed. He told me how he had also bought ten cows. He explained how much milk the cattle could produce. From our conversation, I could tell he was learning about the basics of managing a place. The game made him think about his future too.

This concept—creating your own life—offers an array of factors for us to consider. First, it shows that video games, especially in today's high tech world, hold the allure of an escape for consumers. Want to get away from your job? Anxious to block out a faltering relationship? Just head to Grandfather's old farm. There you can do it all your way and set your own path, far from the stresses of reality. It also reveals the extensive environment these games provide. There are many layers involved, including your business, family life, social life, community, and so on. Lastly, it shows that emotions are just as much at play in these games as the actions themselves.

These components provide a peek into why consumers make purchases, what experiences they like to have, and how they feel during the process. Users might purchase a video game. While they play it, they become interested in upleveling their experience. To do so, they might consider adding features that have a price. These in-game items for sale could help them improve their skills, take out competitors, look a certain way, or satisfy an emotion, such as a desire to be accepted. By participating in these experiences, we can gain a fuller understanding of what drives consumers today—and what we might offer them from our own brand.

As we consider our own role in game play, the possibilities aren't limited to what others are doing. Let's say a car retailer decides to make a game centered on the auto experience. This doesn't have to be a standard racing game. Instead, it could have users go through the steps of manufacturing a car themselves. The game could teach them physics and engineering as they gather the various parts needed. Perhaps they get to offer input on the design. They might test out their ideas along the way.

Users could purchase the auto experience game to help them learn about the manufacturing process. There might be basic features included, such as a car of a certain color and model. Those that want to see a different shade on their vehicle or create a new style could purchase additional features to put their ideas into action. If you incorporate these, you'll want to think about how those transactions take place. Will game players be able to buy what they want while they play? Will they have to leave the game to make a purchase?

Besides the commerce opportunities that building a car game could create, there are brand elements to consider. You could choose to have the game maintain the same logo and motto of your company. It might also provide further information about your products. You could make the game available in different settings. For instance, shoppers visiting an on-ground location might be able to try out the experience for free. They could receive a promotion for playing the game. After spending time immersed in it, they could be asked to share their feedback.

All these activities help to collectively build the brand and keep it in front of consumers. They simultaneously create a seamless experience that allows users to learn about an industry. They also make it easy to move from one mode (like the brick-and-mortar place) to another (such as the online game and virtual brand-related purchases).

We've spent some time looking at the power of video, from engaging with platforms like YouTube to immersing ourselves into the world of play. Perhaps the biggest takeaway to remember about streaming interactions involves the chance to reach customers. Look at these tools as a way to start a conversation. You reach out, a new or existing customer responds, and the relationship takes off. It can blend and mold with time, especially as you post more, monitor the engagement, participate in the community, and provide the audience with relevant, helpful information. You can provide users with games to help them learn and grow. They'll also be able to relate better to your brand and value the experiences they have. If all these efforts ring true with them, you may have found a customer for life.

Perhaps the biggest takeaway to remember about streaming interactions involves the chance to reach customers.

PUTTING IT INTO ACTION

As you think about how to use video and implement video game strategies and concepts, use these questions as a guide:

- ○ Do we currently have videos that we or our ambassadors created?

- ○ What is being said about our brand on video platforms? Is it good or bad? How can it be helpful for our planning?

- ○ What types of experiences do we want to create for video audiences?

- ○ What are our ambassadors doing with video? How can we work with them to build a community?

- ○ How will we monitor and grow our community?

- ○ What can we learn from video games that applies to our brand? How can we put those lessons into action?

CHAPTER 10

Capturing the Long Tail of Commerce

In today's world, you can target a niche market, go global, and overcome cultural challenges to reach more customers and improve profitability.

Let's say you and I meet up to browse the internet together (at this point in the book, I think we may have become friends!). You're looking for a new laptop bag for an upcoming work trip. I suggest a quick search of "laptop bag," which results in seemingly endless results. The listings target almost every customer under the sun.

We look through the search results and verify that you're not a student going to school who needs a yellow backpack with blue stars splashed all over it. You're also not interested in a pink, flowery style that looks a little more like a beach vacation than an office accessory. Secondhand, low cost, high end, fabric, nylon … it's all there. And yet nothing looks right for your next airplane ride and meeting.

Now, I admit, my suggestion was merely to make a point, as I wouldn't usually consider "laptop bag" to be a phrase that would lead you to the ideal computer carrier. We laugh and then I follow up with a few questions regarding what you want. You mention that you're interested in a black leather bag that will hold a fifteen-and-a-half-inch laptop, as your device is that size. We type in "men's black leather bag that will hold a 15.5-inch laptop."

The results? Much better indeed. You sort through a shorter list that contains items that are a perfect—or near perfect—match for you. After reading descriptions and going over the reviews, you make a selection and hit "order." In just a few days, the new large black leather case will arrive. It will be ready for you to take on your upcoming trip.

This exercise led us through an exploration of what is often referred to as long tail commerce. Its inner workings can give us a glimpse into how this method is carried out. We'll spend some time looking a bit deeper into the topic. We'll also review how it has developed alongside e-commerce trends. We'll then discuss how it can be used for your business and as a tool to boost the bottom line.

The Long Tail of Commerce

First presented by Chris Anderson, who at the time was editor of *Wired* magazine, the long tail encompasses the concept of companies selling a smaller quantity of products to a larger number of people. In 2006, Anderson penned the book *The Long Tail: Why the Future of Business Is Selling Less of More*. His thoughts ring true today, especially as the internet makes it easier than ever to distribute goods, save on shelf space, and reach a broader geographic range.[67]

Here's how it works. Items with low sales volumes are spread out on the x-axis of a graph to form a "long tail," which represents the strategy. Even though a small number of each product is sold, there are more product sales overall. This gives companies the chance to increase their profits, provided they are reaching the right people and offering products that fall into a niche.

THE "LONG TAIL"

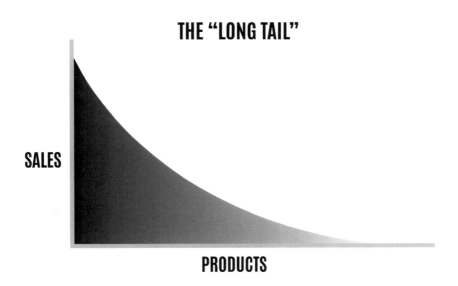

The model differs from others, which might focus on selling large amounts of a small number of everyday products. To see how this everyday concept is often carried out, consider walking through a store. The shelf space might be home to some of the best-selling products, such as the latest gadget, book, or baseball hat for the local team. There are a few models of these, and the shelves are well stocked with them. The retailer expects customers to buy a large amount of the item. Products that fit more into a niche or have a lower demand won't have prioritized positioning or take up the same amount of space as the best-selling items. These low-demand goods might be put off to the side. The store will stock lower numbers of them, as they don't expect many customers to be interested.

Anderson countered that products that have a low sales volume could collectively create a market share that exceeds the relatively few current best sellers. The catch, however, is to make sure the store or distribution methods are ample enough.[68] In a store with limited shelf space, it could be difficult to stock many different low-selling items. Historically, most retailers have prioritized the best-selling pieces and catered to a mass market. In a way, their method is logical. After all, it takes employees time to stock the shelves. Most retailers want to maximize their best-selling goods, making them easy for customers to see and reach.

Enter the online space, and suddenly you don't have as big of a shelving problem. You also might not have as many warehouse issues. You can put up all sorts of different products on your site or online listings. Consider eBay, which oversees the sale of a multitude of product types to customer segments around the world. It works with a plethora of sellers, many of whom put up a few things on the site.

Or you could decide to sell a few niche pieces to a wider audience, as you can now reach customers around the world through your online shop. This creates the opportunity to expand your customer base. You can give people from all over a chance to buy your unique item. In some cases, you might even let them personalize their product for a special touch.

Showing Customers Your Products

Given the trends toward e-commerce, there is great opportunity to make significant profits from products that are difficult for customers to track down. If shoppers can find them at your site, they will be more likely to make a purchase.[69] This is because they know what they're looking for and are ready to buy when they find it (think of your large black leather laptop bag, for instance).

When you add descriptions and keywords to your products that help highlight their features and differentiate them, you're making it easier for search engines to match up your goods with the customers who want them. Peruse

> **Given the trends toward e-commerce, there is great opportunity to make significant profits from products that are difficult for customers to track down.**

Amazon for a few minutes. You'll notice the site's product descriptions include the item's name, color, and sometimes even the size. Rather than "women's coat," for instance, you might see "women's winter oversize coat with faux fur trim, size small." The reviews also highlight details and features of the products.

If you're not familiar with keywords, there are courses on the subject. You can also use tools that are available to help you find keywords that best fit your product line.

USE CASE: STANDING OUT IN THE POOL WORLD

Marcus Sheridan, a highly respected content marketer, spent time early in his career as the owner of a swimming pool company. The company performed poorly and came close to failing. To save it, Sheridan tried marketing more. It took a lot of effort to reach his ideal audience, which consisted of high-end clients.[70]

Before calling the business a failure, Sheridan tried one last approach. He researched his audience and found that most of them had a single question they always asked before buying a pool. Once they had an answer, they were ready to make a decision.

These audience members were looking for input on the cost of a fiberglass pool. Sheridan surveyed his competition and found that not many others were sharing this vital information with customers. He researched long tail keywords. He realized he had to write with descriptive, natural language. Equipped with this knowledge, he created a detailed blog post to answer the most common question his target audience had. He called it "How Much Does a Fiberglass Pool Cost?"

Suddenly potential customers looking for information could find what they needed. They'd also see that Sheridan penned it, which positioned him as an authority in the space. Moreover, they could easily click on his products, which included fiberglass pools. The blog post generated more than $2 million in sales.

It can be easy to think that with the right keywords we'll reach our target audience. However, long tail commerce gets a bit more complex when we take it to a global level. This is because speaking patterns vary from place to place, as do cultures. A product you are marketing in Canada may need different messaging than an item you're trying to sell in Germany (language, for starters!). If you're in India, you might not do well selling beef hamburgers, as the majority of the population avoids eating meat from cows.

To overcome these barriers, you can follow Sheridan's example of marketing pools. Just as he researched his customer base, you can listen to yours. Before marketing to a new country, you'll want to have an idea of what customers within those borders want. You'll try to understand how they like to be treated and how they respond to different messages. Touring the country—if possible—and talking to consultants or other business leaders there could be helpful as well.

As you branch into new markets and regions through your online offerings, there are other doors that could open. For instance, perhaps you are a glassware retailer that caters to European clients. You have online and in-store locations. By looking at your data, you find there is a high demand in France for the wineglasses you're selling on your website. Based on the purchases you see being made from the country, you decide to open a store there. You've never had an in-person shop in France before. Now's your chance to expand your brand further and create more experiences, such as in-store events like seminars and cooking classes.

While there can be challenges related to going global with long tail commerce, it's also true that expanding to new markets is much less expensive than it was a couple of decades ago. Back then, you might have had to set up a brick-and-mortar shop. You would have spent a long time deciphering the language of the region. It could take time to understand the culture and know how to attract clients. You would need to ship goods to your new location and then monitor what happened to them. Did they sell? Did they sit on the shelves? Oftentimes, these steps strained a company's budget and diminished their resources.

In our current world, shoppers on the other side of the world can simply move products on your site into their online cart, carry out the transaction, and wait for the item to be shipped their way. This means you can avoid paying for a physical presence in a new country when you are exploring opportunities there. You could also reduce shipping and warehouse costs, as you may not have to stock inventory in the new country as you start out there.

The Chance to Customize

In the pre-online world, it could get pricey—fast—to personalize merchandise. You might have had to take orders from customers, wait for the material to be produced, and then send it to them. All these steps were far from ideal. Now, however, you can ask customers to give input. Then you can send them exactly what they want.

Perhaps a customer would like a different colored stripe on their new tennis racket. If you give them an option to change the shade, they might jump on the chance to customize their racket. On your end, you won't have to place the order until the customer has selected how they want the product to be customized. You can get it made (ideally close to home and in a vertically integrated way) and sent to the customer in a short time frame. Long tail commerce gives you the chance to better meet your customer preferences.

Including a name on merchandise is one way to add a personalized touch. Pandora Jewelry sells necklaces and bracelets that can bear the letters of your name. Want a Coke with your name on it? Or a Snickers bar with your nickname? Nike shoes with your initials? Almost anything is possible. Moreover, consumers often purchase this name-themed gear for gifts and special occasions.

Long tail commerce gives you the chance to better meet your customer preferences.

To see personalization and customization in action, I invite you to turn to Stitch Fix for a moment. The personal styling company provides clothing for men, women, and children via online channels. When shoppers first sign up to the service, they fill out a survey that conveys their fashion style and how they like to dress. Customers also provide input on the fit they're interested in (loose, formfitting, etc.). They select a range of money they usually spend on clothing. They can add in comments, including if they are looking for a particular outfit for an upcoming event or a note about what colors they never wear. Personal stylists take this information and gather clothing options based on the customer's input.

A selection of clothing is curated for the consumer's "fix," which consists of five items that are mailed to the shopper. In the comfort of their own home, the buyer can try on the different pieces. If they don't care for the item, they simply send it back. If they want to keep it, they make the payment, right from their own home.

There are several takeaways we can gather from Stitch Fix. Notice how the company begins by requesting customer input. Right from the start, consumers can share their interests, preferences, and concerns. The stylist is there to help. Even though they won't meet in person, a relationship can form. After receiving their five items, the customer can provide feedback to the stylist. The shopper can also choose when they would like their next shipment, which could be every two to three weeks, once a month, every other month, or once every three months.

Before the next shipment, the stylist will review the shopper's comments. The curator might even reach out to see if the customer is interested in trying something new or if they want to stick with what they know and like. This give-and-take allows the consumer to customize their shopping experience. They get help picking out clothes. Throughout the process, they might discover new features and styles that they like. Conveniently, they can make purchases right from the comfort of their own home.

Crunching the Numbers

While it can be appealing to think of selling products to a niche market, we do have to be financially attentive to make sure we break even (and hopefully do better than that!). For each product, there is typically a minimum threshold related to expenses that will need to be passed for it to be cost effective. Perhaps you need to mail your products. There will be a shipping cost, duties to be paid, and potential warehousing costs. All these steps will have to be covered before you even make one dollar in sales.

For this reason, the long tail of commerce focuses on the costs needed to produce and distribute your niche products. Once you create an estimate for the minimum cost to produce an item, you can think about how many of those items will need to be sold to break even. Going one step further, you'll want to consider how many products you need to sell to receive the desired profit.

When running the numbers, you don't have to shift your strategy to only long tail commerce. Instead, you might find you can market some popular, mainstream products to a mass audience. You might also be able to sell low-demand products to a broad global audience. You don't have to choose between traditional approaches and long tail commerce. The two can complement each other and be an integral part of your overall strategy.

PUTTING IT INTO ACTION

As you evaluate how long tail commerce could work in your reimagined business model, discuss the following questions with others on your team:

- O Do we currently sell to mainstream customers or a niche market?

- O How can we expand to reach an overall greater customer segment?

- O What opportunities do we see to offer more customized products to shoppers?

- O How can the long tail commerce approach be used in our organization? How would it complement or impact our current strategy?

- O What are the costs and benefits related to our ideas for long tail commerce?

CHAPTER 11

Data and Increasing Margins

Capture information the right way to react quickly, make decisions faster, and move the needle toward higher profitability.

Have you ever walked past a store display of merchandise and seen a sign over the items stating in bold letters, SALE – 25% OFF? And then strolled by the same setup a week later to see a slightly revised version of the discount, offering customers a 30 percent markdown? What happens two weeks later when the third sign promises 50% OFF? Finally, the poor remaining inventory gets the CLEARANCE banner over it. The retailer hopes that customers who previously passed by (like you) or newcomers will take notice and buy …

though at this point, as we know, the retailer is hardly making a profit from the sale.

Quite the opposite. The store may be trying to free up shelf space so new items can be put into place, but the next order won't resemble the shipment that refused to sell and had to go through several rounds of price cuts to get into shoppers' carts. Steep discounts may provide a thrill to thrifty consumers who love sifting through mounds of merchandise. For most retailers, it's a sign of an order that didn't perform as planned. In worst-case scenarios, poorly selling items mean that budgets must be revised, revenues adjusted, and financial statements altered.

Having to sell items at a loss is just another reason why companies can run the risk of bankruptcy. It's also an act that can generally be avoided, provided systems are in place to capture data. The information can flow to the right places and land on the screens of those who make decisions. You see, a stack of merchandise, such as an order of one thousand black laptop bags, might seem like a great idea when the order is placed. However, if those thousand computer briefcases must first be manufactured at a facility on the other side of the world and then shipped to the right stores, the calendar might march on over several months. During that time, customers could decide they prefer to buy brown laptop bags, not nighttime-toned fabric. For the retailer that put in the order and stocked it in stores after the trend changed, the next step often consists of lagging sales. This is followed by discounts designed to get rid of the inventory. Of course, the strategy cuts into profit margins and can lead to uncertainty over future orders.

In cases such as this, there is often a disconnect in data. There could be a gap in information between the supply chain and the brick-and-mortar store, for instance. After the brand orders the thousand sets of computer bags, executives and managers might not know what happens to the shipment once it hits the store shelves. Only months later, when it is still sitting inside the shop's walls, does someone notice that the items have been there a long time. By this point, it is too late to make many changes. The main objective then becomes getting the shipment out the door. The brand loses money in the process.

This type of scenario often plays out when departments are established and not integrated together. In other words, the business functions get set up, but no one connects the dots between them. You might have a marketing department, a sales function, a procurement branch, and so on. Marketing might not know at a detailed level what is going on in sales, and vice versa. Procurement personnel may not meet with marketing or sales on a regular basis. As such, they don't have a clue about what's going on in those departments.

These disconnects had a logical explanation in the past. Turn the clock back ten or twenty years, and you'll see (or recall) that implementing software such as an enterprise resource planning (ERP) solution to tie these functions together was a painful, slow, and expensive process. It was so clunky, in fact, that it sometimes took an organization two years to fully implement the new system. At every step along the way, costs were racked up and morale took a dip (or a dive!).

Fast-forward to today. A quick search will help you see that putting in software to integrate business departments doesn't have to cost an arm and a leg. Better yet, it won't take years to set up. Technology advances provide additional benefits, including that you can pick and choose which integrations you want. You don't have to opt for a full-blown ERP project. Instead, you can choose a few pieces of software that will help you capture data that will be meaningful to your company and its operations.

Making this switch doesn't start with buying the software. Rather, it begins by looking at data in a new way. Think about information that would be relevant and how it could help the company. What trends would be meaningful to track? What information would be valuable to see in real time? How could data help you make better decisions about orders and sales and everything in between?

With a few solutions, you might be able to spot that those one thousand black computer bags haven't moved right away. Better yet, maybe you never even order those money-draining goods. Let's spend some time looking at what integrated systems can do. We'll observe how they can be used to move the needle away from losses and toward higher profitability.

Ordering the Right Way

Let's relive the case of the sad loss of the thousand black laptop cases. What if, instead of ordering one thousand of the same-colored items right away, the retailer decided to order ten black laptop bags, ten brown ones, and ten cream-toned options? If an integration system is in place, the retailer could watch how each color sells. Perhaps the system identifies that all the brown ones fly off the shelves. It reveals that the black and cream-toned bags lag in sales. As a result, a larger order for brown bags might be placed.

There's one other component that must be in place for this scenario to play out successfully. The retailer will want to cut out long manufacturing times, which often occur when items are assembled offshore or outsourced. Instead, the company might choose to have its own manufacturing facility within miles of its store locations. This makes it easy to order, fulfill demand, and make changes without having to wait for months to see results after decisions have been made.

Perhaps the biggest plus of this arrangement—integrated systems to display real-time, valuable information coupled with close-to-home manufacturing—lies in the opportunity to protect margins. When a retailer can easily see what the customer wants, they can provide it in a timely way and meet the demand in the market. Merchandise moves through the store (or gets put in online shopping carts). Items don't have to be discounted. This allows for a better return on investment and positive financial statements that reflect this flexible system.

USE CASE: CROWNING ZARA: A REIGNING RETAILER IN THE WORLD OF FAST FASHION

As the flagship brand of Inditex, Spanish clothing giant Zara knows fast fashion—and profitability.[71] How speedy is it? New styles are prototyped in just five days, and the entire design process can typically be completed in just fifteen days. Local labor is used. Its clothes are manufactured right in Spain, creating a vertically integrated system that helps operations run efficiently.

Zara began as a brick-and-mortar shop in 1975 in the main shopping area in A Coruña, Spain. It featured cheaper look-alike designs of famous and more costly dress styles. Hitting success, more stores opened throughout Spain. Today the chain has more than two thousand locations in various countries.

Though it has grown, the brand maintains its focus on quick, agile production with a nod to the ever-changing fashion trends. During its early years, while competitors produced just a few thousand SKUs a year, Zara cranked out several hundreds of thousands of SKUs annually. These all varied in color, size, and fabric. To create them, Zara's designers watched the catwalk displays for each season and then made replicas that would work for the mass market.[72]

Customers loved the concept, and they also reveled in the chance to go to Zara's stores, where they could walk into a beautifully designed, upscale-feeling store full of quickly changing merchandise that met their interests. (Remember the Avenue des Champs-Élysées in Paris, with all the luxury stores? Guess what? Zara has a place there too.) The fast-moving, tightly numbered inventory did more than allow the store to adjust to changing demand. It also created a sense of scarcity.

Shoppers couldn't be guaranteed they would find the same item at the same store (or online) in the weeks ahead, making it that much more appealing to purchase right away—and then wear the copycat runway deals to the office for work, to the beach during holidays, and everywhere in between.

The rage continues, making Zara a standout example of the power of fast fashion backed by an integrated system that allows for changing trends and demand. With headlines such as "Zara's Spring Collection Is Full of This Season's Biggest Trends"[73] and "I'm Going to Zara for All My 2022 Outfits,"[74] along with "These 13 New Arrivals at Zara Are All Winners in Our Book,"[75] the company has shown it can weather a pandemic and e-commerce trends and come out ahead. Companies looking to do the same can take note of its nearby manufacturing facilities, connected systems to follow trends, and efficient processes to adjust as quickly as women's fashion (known to pivot on a whim) does.

Show Me the (Right!) Data

The issues related to integrating systems and leveraging data don't stem from a lack of information. In today's world, data abounds. We've looked at several instances of data mounds already, including the details that Facebook and mobile phones track on us every day. The problem, then, lies in what to do with all that data. While there are clues in the information that is gathered about merchandise and sales, it is a bit like finding a pen in a pile of snow. It can be done, but the process will be long, tedious, and time-consuming. Few will want to succumb to the burdensome effort. Moreover, this practice is a drain on valuable resources, including labor and time, which can be costly for firms of any size.

What we need is a way to capture the data and integrate systems in the right way. A data warehouse that can be accessed to run analytics is definitely a step in the right direction. To understand where your organization is on its data journey, think about the following: Do we have systems that are capable of talking to each other? Do we have a way to follow the trends we want to see? Are we being notified when an item is selling fast? Do we have a place where key metrics can be seen at a glance? Is there a way to produce regular reports that have meaningful information? Are we able to use the data to make decisions quickly and implement them on a timely basis?

A smooth-running system will provide relevant data at your fingertips ... and you won't have to pay through the nose for it. Consider a dashboard that you can open on a screen. It shows you what is selling and where it is selling, whether that be online or at a store location. You could have this information displayed in real time (or as close to real time as possible). You should be able to drill down and see details. At what time did this pair of white jeans sell? From what location? How many of these white jeans have sold in the last week? What about the last month? How many pairs of white jeans are left in stock at a specific location?

Think about how your decision-making could be impacted if you could follow a path such as Spring Collection >> Jeans >> Women's Jeans >> White >> Size 28 >> Galeria Location, Houston, Texas. Suddenly it is as if you were standing in the retail store, counting the number of pairs left hanging on the rack in the front of the store and checking to see if there are more in the back. Instead, you're looking at a screen that shows you everything with just a few clicks of a button. Better yet, you can visit any location you want, courtesy of the dashboard, and do a deep dive.

Data can help us look ahead, especially when it comes to production planning. Say you have a system that shows you data related to browsing. This information will give you key insight into what consumers are searching. They might not buy anything, but the fact that they are looking means they are interested in the item. Perhaps you see that high numbers of people are searching for "colored patio lights." This could lead you to decide to produce patio lights in different colors.

In addition to going under the layers to get an inside look at what's going on, you can zoom out to see the big picture. You might look at what's selling overall. You could check what is getting picked up by customers. This can be particularly helpful, as it gives a real look at how the demand is shifting and provides guidance on what to order next.

Imagine if you had a way to spot the best sellers without having to wait for a quarterly report that looks back at the previous months. What if you could open the dashboard and check for flashing lights? These might be alerts letting you know that a product is selling extremely well. Perhaps one item in your inventory, such as a kitchen cutting knife, was mentioned on a cooking show the night before. After getting the chef's recommendation, TV viewers are pulling out their phones and laptops and looking to make orders. If you're unaware of this quick uptick in sales, you could rapidly run out of stock and lose potential sales. If you spot the increase in demand, you can order more right away and replenish the supplies to give customers what they want—no discount needed! In the same way, an alert on a dashboard could notify you of merchandise that is not moving. Remember the lonely thousand black laptop bags? What if you had a light on your dashboard that flashed early on when they weren't selling? Perhaps you could rethink your marketing strategy, change an upcoming order, or look for a way to package the computer luggage with something else to make it more attractive to customers. The possibilities expand when you have the right information at the right time and in the right place.

The Underrated Preorder Function

Do you know if your company has a preorder feature? Many websites today come with this button. It's not typically a question of being able to afford this function. What we need to do is think about how we want to use it. We can consider how we can put it to work toward future revenue and increased customer satisfaction.

Here's what can happen: There might be a preorder feature on a website. If it's not prominently placed, customers probably won't see it. Even if they do view it, they might be unsure of what they can preorder, when it will become available, and why they should commit.

The possibilities expand when you have the right information at the right time and in the right place.

All of this gets solved when we make the most of the preorder function. Certainly, we can have it available for items that are currently out of stock. The customer doesn't see the printer they want? No problem. They can select the preorder and reserve a printer from the next batch that comes in. You can make it so the customer can track the printer they want and see when it will be delivered. They can then rate their experience after it arrives and offer feedback for future orders.

Now let's take it one step further. Perhaps the customer wants the printer. They would also like a couple of additional features put on it. Are they able to provide this commentary as part of the preordering process? Is there a way for them to communicate that they like the printer but would prefer to order it in a different color? What if they want a tone that isn't currently available or being manufactured?

The preorder feature can help you drum up interest in future products. It also allows customers to give input regarding what they would like to see. When they feel like they can help design a product or have their voices heard, they'll pay attention. And they'll be ready to buy and let others know that their ideas got put into action. Again, here you might not need to offer a discount. After all, the consumers have access to what they want and feel valued. Chances are, they're willing to pay the price you set to coincide with this experience.

THINKING LIKE TESLA

Before electric cars came on the scene, many manufacturers felt there was no opportunity in vehicles that didn't run on gas. They knew that in the past, it had always been too expensive to make an auto that would be based on electricity. They were also aware that technologies to put such a car together hadn't existed several decades ago.

Then Elon Musk decided to check on how those processes could operate, if given the chance. He found that prices related to production had come down, technologies had advanced, and electricity options were at a point that, with a few plans and testing—voilà—an electric vehicle could drive around. To finance the production of vehicles, Tesla predominantly used the preorder function. Customers could request a vehicle before it rolled out of the manufacturing plant. Through this arrangement, Tesla created a long list of customers.

In addition to financing, Tesla found another benefit from the preorder function. Shoppers were able to specify which type of car they preferred, including the color of the vehicle. Tesla used this information to better estimate how many vehicles to make.

When fuel prices increased in 2022, consumers responded, and Tesla's sales went through the roof. By March, the company's vehicles were sold out for the year. Despite supply chain issues, the company continued to experience record growth.

What did Elon Musk do that was different from other car manufacturers? He decided to check for himself to see what was available in terms of technology, solutions, and costs. The same lessons can apply to data and how it can be used in business processes. Rather than deciding it is too costly and cumbersome for us to figure out, we can think about how our organization could benefit from customized insights and real-time reporting. Once we start to picture our business differently, we can see how data could play a role. Then we can search for solutions that will fit our specific needs and help us move the needle to higher margins.

Just as we said, the customer is at the center of everything, so data needs to sit in the middle of our processes. When this happens, data will capture what the customers have done, what they are doing, and what they will do in the coming months. And that, my friends, is where the magic starts and the profits grow.

> **The customer is at the center of everything, so data needs to sit in the middle of our processes.**

PUTTING IT INTO ACTION

As you contemplate the use of data and how to leverage information, answer the following questions:

- How do our current systems interact?

- Do we have gaps in how information is shared?

- What sort of data would be valuable to know to help us make decisions?

- Which of our departments tend to be fragmented and could benefit from integration?

- How does our current decision-making process take place? How could it be improved?

- Who in our organization would benefit from having relevant data that is easy to share?

- If we didn't have to think about budgets, how would we like to use data in our organization?

- What types of solutions and integrations would we like to see? How can we make those visions a reality?

CHAPTER 12

Building Your Toolbox: A Set of Best Practices

To forge ahead into the coming years, set up your people and systems the right way, then bring in the digital functions to support your goals and allow your dreams to unfold.

Let's fast-forward five years and imagine we are sitting in the boardroom of your company. You've asked me to come in (thanks, by the way, for the invitation!) and help you evaluate the organization. You're looking for feedback on how systems are working, the overall mindset of the people, and, of course, the financials of the company.

In this setting, as you pull up charts, graphs, and financial statements and I sit at the table absorbing it all, what do you think we are going to talk about? Will we be lamenting the recent losses? Will we be comparing your firm's performance to competitors who seem better equipped to move forward in the commerce world and its complex online and brick-and-mortar trends? Will you be asking for advice on how to change the mindsets of executives and personnel and start thinking outside the box?

Of course, we can't see into the future for certain. Still, let's continue with this pretend scene for a few moments and switch it up. What if, instead of worrying about what to tell shareholders in the coming reports, you and I were documenting strategies you've implemented? Perhaps you've been asked for details on your processes, which now serve as an example for others to follow. What if your company was leading the way in its industry and had a strong grasp on the interconnectivity of virtual and on-the-ground activities? What if you were showing me a long list of best practices that you started five years ago and that had already made a big impact on your profitability?

I'm guessing you're leaning toward the latter option (me too!). The good news is that I've spent time studying brands and retailers. I've watched trends and evaluated digital tools. I'm well in tune with what these best practices could look like. I can also assure you that they don't have to break the bank.

The first step in long-term planning does not necessarily require an increase in spending. Instead, it focuses on making the most of your organization's talent base and leveraging their strengths. It also includes coordinating these best practices to align with your company's overall goals. The best part: It's there and waiting for those who dare to dream. Let's open our minds and think about what to include in our toolboxes for a better tomorrow.

Set Up Your Teams

As we've seen in previous chapters, digital transformation doesn't begin with selecting a digital tool and implementing it, choosing another one, and so on. Rather, it starts with making sure the way we do things is set up to meet (and foresee) consumer preferences, to build relationships and cater to customers, to align our brand's online and offline presences (and make the most of them), and to capture data in a relevant, meaningful way. Behind all these systems is a powerful force: our staff. While our workforce can bring individual strengths to the table, there is synergy that stems from well-designed groups that are built to work in tandem with one another. Here are a few key teams to consider setting up in your organization to face the future.

THE CUSTOMER EXPERIENCE GROUP

How do consumers find your brand online? How can they access your products via the internet? Are there ways to purchase products from other devices, such as the TV, fridge, car, or other everyday equipment? Are headsets available that they can access or use in their homes to better understand your product? All these questions can be presented to the customer experience group. Their role will be to zoom in on the customer and understand their paths to purchasing.

While it's important to know how customers first connect with a brand, it's just as vital to see how their experience plays out over time. For instance, what happens after

> **While our workforce can bring individual strengths to the table, there is synergy that stems from well-designed groups that are built to work in tandem with one another.**

they buy an item? Are they contacted again? How do they receive information about upcoming promotions? How do they stay in touch with the brand?

One way to approach this, as we've observed previously, is to step inside the customer's shoes. Encourage team members to do exactly this. View their role as a fluid one. By that I mean we'll want to avoid certain gaps, such as spending a day looking at how customers find our brand and then closing the door on the task for another year, at which time we revisit how consumers can locate us. Instead, an ongoing approach involves studying consumer trends, looking at what's happening in the market, evaluating current systems to see if they fit customer needs, and envisioning what could be in the future.

This team will also recognize that the framework that expects customers to get into their car, drive to a store, look for and purchase an item, and go home is long in the past. Rather, their journeys wind in and out of the mobile world. They have touch points at stores and pop-up shops. The path includes interactions and purchases from— literally—anywhere. For these reasons, mapping out the customer experience means looking at the buyer journey in a way that is very different from what was done a couple of decades ago. Based on this recognition, team members will work to make sure the entire experience is as seamless as possible. Their goal is to make sure that paying for an item—and receiving it—is so easy that customers will want to do it again.

THE VOICE OF THE CUSTOMER GROUP

Do customers want to receive text messages? Do they enjoy a video call after their purchase? What do they think about the headset they were sent—did it impact their decision?

These questions will be presented to the group that focuses on the voice of the customer. Their goal is to listen to customers and take a deep dive to see what they want and how they prefer to be treated. The players on this team might seek to have conversations and build relationships with repeat clientele.

This personal emphasis can go a long way. It helps the organization determine what strategies could work well. It also serves to help clients feel appreciated and valued. Showing them respect typically works in an organization's favor, as these shoppers are likely to turn to the same brand for their next purchase.

AMBASSADORS, WE LOVE YOU

What do your ambassadors want to see in five years? If you're not sure, just reach out and ask. Chances are, they'll be anxious to share their thoughts—and will be full of ideas. When they express them, listen … and take notes. The advantage here is that ambassadors are well informed of your brand, and they like it.

Moreover, they want to share it with others. This motivation can result in free marketing for your brand. There's no need to spend thousands of dollars on a campaign if you have thousands of ambassadors who freely share your latest product or offer with their audiences. Think of musicians and their fan clubs. These rock-star lovers are passionate about the beat of their beloved performers and follow them closely. When musicians interact with their fan base, those admirers feel valued and inspired. They'll attend more concerts and follow a band's social media pages. They'll also buy the next album when it's released. They might even preorder it if they have the chance or listen to songs before they are released to the public. When that album hits the market, those same fans will be quick to share what they think about it with others. Consider the possibilities of having ten thousand die-hard fans posting about a newly released album. Talk about great advertising!

How do you make sure you're getting the most from your ambassadors? Perhaps an initial evaluation of the ambassador group and how it is being used could be this team's first objective. Team members could then look for ways to better communicate with ambassadors. For instance, could people who love your brand be invited to attend an exclusive webinar? Would they like to stay in touch via email? Are they interested in receiving free merchandise, and if so, what? How could they help promote an upcoming product? Sorting through these possibilities can help you build a community—and then benefit from your efforts.

THE DATA FLOW GROUP

Is data in the middle of your processes? Is it scattered throughout them or pushed off to the side? Knowing where data stands in an organization is often the first task for a data flow group. This team has an important job at hand. That's because everything about data, from what is captured to how it is used, is becoming increasingly essential for all organizations. As we saw in the previous chapter, we can take steps to make sure it is accessible and being used to help make decisions.

If you have team members in this group who are excited about data, they can help your organization understand where it currently stands and how changes could be implemented. They don't have to stop there. This group can continue to review and improve the ways data is gathered and managed. They might be able to take your organization from one that is lacking in data knowledge to one that is leveraging the power of the right data. When that happens, you can expect to see higher profit margins and positive returns on investments.

THE IDEA GROUP

I sometimes jokingly refer to this team as the geek group, but it is with an acknowledgment that these individuals have incredible, creative minds and bring value to the table. Organizations will do well to pay attention to them. This group should be allowed to spend time coming up with crazy ideas about how the business should change. Its members will look at new gadgets, programs, and tools on the market. They will evaluate how these solutions could be used in the company.

A solid creative group will likely come up with an ongoing flow of ideas. The key to drawing benefits from them is to evaluate what can be valuable to the firm—and what should be passed on. Most of the ideas from this type of group will typically not be usable or applicable, and that's okay. If even 10 percent of their innovations are implemented, it could make a big difference in the overall performance of the company.

Before Google went public, its cofounders, Larry Page and Sergey Brin, wrote, "We encourage our employees, in addition to their regular projects, to spend 20 percent of their time working on what they think will most benefit Google. This empowers them to be more creative and innovative. Many of our significant advances have happened in this manner." Case in point: a few of the breakthroughs that came from this approach include Gmail, AdSense, and Google News.[78]

Finding the Right Players

Great teams are only as strong as their members. In every group, we'll want to strive to have people who are curious and interested in participating. We'll make sure that the members love what they are doing and that they are passionate about the team's focus. If it is a burden for these individuals to participate, it may be a sign that they'll be better off in a different group—or not in a group at all.

How do we form these teams? The process doesn't have to be complicated. It might start by surveying employees and identifying those who have interests and seem to stand out above the rest in terms of energy, commitment, and leadership. It may involve sending out a survey to staff members and asking if they'd like to participate in one of the teams. Those interested in joining a group can mark which one they'd prefer. Another way might involve carrying out interviews with employees to learn about their strengths and then placing them in the appropriate group.

Keep in mind as you organize teams that the members may naturally come from within the organization. That said, groups could include outside members as well. A brand ambassador might not be an employee, for instance. Or you could decide you'd benefit from an outside perspective in a certain area, such as having a consultant come in and map out customer experiences. Third-party involvement, provided the environment is suited for collaboration, can bring additional benefits. These outsiders could help mentor and instruct employees of the organization, pass on new skills, and broaden the perspective and mindset of the group.

When bringing in members to a team, questions will arise regarding the amount of time and dedication required. For instance, will the employee spend all their working hours within the group? Will they commit to putting in several hours each day or each month? How will they divide their time between other tasks and the group's focus?

The example of Google helps us sort through some of these questions. In its case, we saw that employees spend 20 percent of their time on creative projects. If a team member works forty hours a week, that comes to eight hours of time (one full day or an hour or two each workday) used toward thinking of new solutions. Be careful not to view these periods as lost time. Instead, the hours spent sorting through opportunities and highlighting options are vital to providing the innovation and framework to drive the company forward. This group might come up with new solutions, optimize existing processes, and open the doors to breakthrough products with great potential.

Take Away the Budget

Isn't this a chapter on best practices? you may be thinking. You are correct. Allow me to explain what we are referring to when we talk about x-ing out the budget. The idea isn't to spend without parameters or to operate without accountability. Rather, the concept involves turning our heads away from the numbers and giving ourselves permission to think outside the box. (We don't start spending at this point, so there's no need to worry about blowing the budget and having to make up for it next year!)

Here is what I mean. Consider this statement: "There is a twenty-thousand-dollar budget for computers." It probably makes us think, at first glance, that we must find computers that meet the organization's need and fall into this price range. What if, however, we didn't run out to search for equipment and instead asked a counterquestion? What if we responded, "Do we need computers?"

Maybe the organization does, but the point is to think about how we want to do business. For instance, what if we talked to some of our groups and gathered feedback from them on how we could best serve customers? Maybe we find out that customers are really interested in a video shopping assistant. Perhaps our customer voice group has carried out extensive research and talked to many repeat clients. They suggest we make this possible so that a customer can be online or in a store and easily access a brand representative who can assist them. There's just one problem: we don't have "video shopping equipment" listed on any line in the budget.

This is exactly why great things can happen when we take away the budget. When we begin from the other end, starting with what we want for our business, we can work backward and see how much we need to make it happen. In the case of the computers and the video shopping assistant system, we might decide to evaluate the two. Should one be prioritized? Can we do both? What about a board meeting to see how we want to place our funds, based on the company's way of carrying out business? When this happens, we can find budget solutions that help drive the organization forward. Maybe we are unable to implement everything we want at once, but we will likely be able to make a start. After that, we can lay out a path going forward and think about how to allocate resources in the future.

Testing, Testing for the Win

When we have great ideas and new ways to use resources, we don't have to completely revamp the budget and overhaul our entire system. If we take this route, in fact, we could be in danger of allocating too much, too soon. Instead, I tend to suggest a testing strategy that involves pilot projects and small samples to try out a new product or solution. If it goes well, we can think bigger. If it doesn't, well, we haven't overextended our resources and put our budgets (and company) at financial risk.

When we test out our strategies in an honest, transparent way, shareholders notice. And they don't necessarily condemn the losses. Looking at Amazon today, which brought in nearly

> When we have great ideas and new ways to use resources, we don't have to completely revamp the budget and overhaul our entire system.

$470 billion in revenue during 2021, we might be quick to say that the innovation was a smart one. However, turn the clock back to 2009, and you'll see the company reported less than $25,000 in revenue.[79] Rewind a bit further, and it becomes apparent that Amazon lost a combined $2.8 billion during its first seventeen quarters as a public company.[80] To its credit, the company remained up front with investors and shared those losses from one quarter to the next. Not all shareholders were quick to punish the company—and those who hung on to what started as an internet retail experience and grew to become an online retail behemoth benefited in later years.

Amazon hung on, and we can too. Those pilot projects can lead to exciting opportunities, especially when we put our teams to work and use their ideas to fuel those endeavors. When this happens, an entire organization can transform. Employees see that they're adding value, customers feel appreciated, and innovations have the chance to thrive. When done well, the future is full of positive financial statements, an enthused workforce, and new possibilities to explore.

PUTTING IT INTO ACTION

As you think about setting up best practices in your organization, use the following questions as your guide:

- Do we currently have any teams in place?

- Which teams would we like to create? Do we want to prioritize some over others?

- What strategies would work well for our organization to find team members? Do we want to stay inside or look for outside support?

- How are we connecting with our ambassadors, and how can we better build a community?

- How do we want to do business in the coming years?

- What tools are needed to get us there?

CONCLUSION

At the beginning of the book, I invited you to dream. So now, as we come to the end of our time together, I'd like to close with a question: Were you able to dream?

From my experience, I've watched organizations and leaders struggle with words like imagine, play, and experiment. These terms aren't always associated with the business world, and it can be hard to strike a balance between productive agendas and downtime. It's often much easier to focus discussions on numbers, reports, and projections.

This, however, is the very reason I wrote the book. I've seen that companies do best when they tear away from the budget meetings and long to-do lists. Certainly, these are important and have their place. Let us remember, however, that retail is changing. Those who don't take the time to reenvision how their company interacts with technology will pay a dear price in the not-too-distant future. The same holds true for organizations that don't give flexibility to their

THE GREAT DIGITAL TRANSFORMATION

workforce, put the customer at the center of everything, or track data in a meaningful way.

The exact opposite awaits those who tap into their inner child—if only for a few hours. I advise you to open your mind, take away all financial barriers and limitations, and create an image that features happy customers and high profit margins. Consider why the consumers are satisfied and how your company could better engage with them. And, of course, don't forget about the profits. As we've seen, there are often ways to prioritize changes and even implement revenue-generating models.

> I've seen that companies do best when they tear away from the budget meetings and long to-do lists.

Our time together on these pages is coming to an end, but the chance to dream will always be there. If you're looking for help to brainstorm new ideas and make manageable, profit-producing changes, my company specializes in guiding organizations through their transformation. I invite you to connect with me. Let's dare to dream and then enjoy the journey.

ABOUT THE AUTHOR

Gerard (Gerry) Szatvanyi is a founding member and CEO of OSF Digital, a global commerce and digital transformation leader with expertise in connecting technology and strategy to drive business goals.

Gerry has more than twenty years of experience growing start-ups and medium-sized tech-enabled businesses and driving them to peak performance. With a background in cloud applications, software integration services, and consultancy, Gerry's impressive client and business portfolio sets him in the new breed of global tech entrepreneurship. He holds management and board positions in several tech-enabled businesses. Gerry has an MS from Laval University, Quebec City.

Gerry is a founder of the OSF Foundation, which contributes to various charitable, cultural, and educational causes in the communities where OSF Digital operates. OSF Digital is also a proud member of the Pledge 1% Community.

CONTACT

Gerry Szatvanyi is the CEO of OSF Digital, a leading global commerce and digital transformation partner to the world's biggest brands. With expert status in B2C and B2B commerce and several Salesforce awards for multi-cloud innovation, OSF Digital seamlessly guides enterprises through their entire digital transformation journey. With customers in various industries around the globe, OSF Digital provides personal attention and the highest level of connection with a local presence throughout North America, Latin America, APAC, and EMEA.

To connect with Gerry Szatvanyi, follow him on LinkedIn at **https://www.linkedin.com/in/gerardszatvanyi/.**

https://www.gerardszatvanyi.com/

To engage with OSF Digital, and to learn more about OSF Digital and its innovative products and services, please visit:

Website: **https://osf.digital**

LinkedIn: **https://www.linkedin.com/company/osf-digital/**

Twitter: **https://twitter.com/osfdigital**

Facebook: **https://www.facebook.com/OSFDigital/**

YouTube: **https://www.youtube.com/c/OSFDigital**

ENDNOTES

1   Anne D'Innoncenzio, "Tuesday Morning Seeks Bankruptcy Protection, 5th Major Retailer to File Chapter 11," *Chicago Tribune*, https://www.chicagotribune.com/coronavirus/ct-nw-coronavirus-tuesday-morning-bankrupt-20200527-fduq7zyxuvdppcn3cfl-4lqud4e-story.html.

2   Retail Dive Team, "The Running List of 2021 Retail Bankruptcies," Retail Dive, June 11, 2021, https://www.retaildive.com/news/the-running-list-of-2021-retail-bankruptcies/594891.

3   Hayley Peterson, "A Tsunami of Store Closings Is about to Hit the US—and It's Expected to Eclipse the Retail Carnage of 2017," Business Insider, January 1, 2018, https://www.businessinsider.com/store-closures-in-2018-will-eclipse-2017-2018-1.

4   Kate Taylor, "One Statistic Shows How Much Amazon Could Dominate the Future of Retail," Business Insider Africa, November 1, 2017, https://africa.businessinsider.com/

strategy-one-statistic-shows-how-much-amazon-could-dominate-the-future-of-retail-amzn/hn9x3jr.

5   Derek Thompson, "What in the World Is Causing the Retail Meltdown of 2017?," *The Atlantic*, April 10, 2017, https://www.theatlantic.com/business/archive/2017/04/retail-meltdown-of-2017/522384.

6   Society for Human Resource Management, "Interactive Chart: How Historic Has the Great Resignation Been?," March 9, 2022, https://www.shrm.org/resourcesandtools/hr-topics/talent-acquisition/pages/interactive-quits-level-by-year.aspx.

7   Kim Parker and Juliana Menasce Horowitz, "Majority of Workers Who Quit a Job in 2021 Cite Low Pay, No Opportunities for Advancement, Feeling Disrespected," Pew Research Center, March 9, 2022, https://www.pewresearch.org/fact-tank/2022/03/09/majority-of-workers-who-quit-a-job-in-2021-cite-low-pay-no-opportunities-for-advancement-feeling-disrespected.

8   Melanie Langsett, "Elevating the Workforce Experience: The Work Relationship," Deloitte, March 9, 2021, https://www2.deloitte.com/us/en/blog/human-capital-blog/2021/the-value-of-meaningful-work-to-workers.html.

9   Melanie Langsett, "Elevating the Workforce Experience."

10   Brian Stone, "3 Things Employees Want Most in 2022," TechRepublic, February 3, 2022, https://www.techrepublic.com/article/3-things-employees-want-most-in-2022.

11   Brian Stone, "3 Things Employees Want Most in 2022."

12   Society for Human Resource Management, "Managing for Employee Retention," 2022, https://www.shrm.org/ResourcesAndTools/tools-and-samples/toolkits/Pages/default.aspx.

13   Alison Green, "My Boss Wants Us on Zoom All Day Long," *New York*, September 22, 2020, https://www.thecut.com/article/my-boss-wants-us-on-zoom-all-day.html.

14   Apollo Technical, "Statistics on Remote Workers That Will Surprise You (2022)," January 16, 2022, https://www.apollotechnical.com/statistics-on-remote-workers.

15   Adam Grant, "The Real Meaning of Freedom at Work," *The Wall Street Journal*, October 8, 2021, https://www.wsj.com/articles/the-real-meaning-of-freedom-at-work-11633704877.

16   Brian Kropp and Emily Rose McRae, "11 Trends That Will Shape Work in 2022 and Beyond," *Harvard Business Review*, January 13, 2022, https://hbr.org/2022/01/11-trends-that-will-shape-work-in-2022-and-beyond.

17   Brian Kropp and Emily Rose McRae, "11 Trends That Will Shape Work in 2022 and Beyond."

18   World Population Review, "Obesity Rate by State 2022," 2022, https://worldpopulationreview.com/state-rankings/obesity-rate-by-state.

19   Brian Kropp and Emily Rose McRae, "11 Trends That Will Shape Work in 2022 and Beyond."

20   Primark Careers, "Your Day, Your Way," 2022, https://careers.primark.com/your-day-your-way.

21   Ross Freedman, "Consumers Are Hungry for an Experience-Based Connection with Your Brand," *Forbes*, November 21, 2019, https://www.forbes.com/sites/forbesbusinesscouncil/2019/11/21/consumers-are-hungry-for-an-experience-based-connection-with-your-brand/?sh=2f40677658fe.

22 Matt Wujciak, "How Nike Is Enhancing Future Customer Experience through Brand Image," Customer Contact Week, April 28, 2020, https://www.customercontactweekdigital.com/customer-experience/articles/how-nike-is-enhancing-future-customer-experience-through-brand-image.

23 Thomas H. Davenport and Nitin Mittal, "How CEOs Can Lead a Data-Driven Culture," *Harvard Business Review*, March 23, 2020, https://hbr.org/2020/03/how-ceos-can-lead-a-data-driven-culture.

24 Nike, "Nike. Just Do It," accessed August 12, 2022, www.nike.com.

25 Iterable, "Nike's Customer Experience Puts Their Best Foot Forward," August 11, 2021, https://iterable.com/blog/nikes-customer-experience-puts-their-best-foot-forward.

26 WBR Insights, "Here's How Nike Is Turning Data into Unrivaled Customer Experiences," NGCX, 2022, https://nextgencx.wbresearch.com/blog/nike-data-unrivaled-customer-experiences-strategy.

27 WBR Insights, "Here's How Nike Is Turning Data into Unrivaled Customer Experiences."

28 Bruce Y. Lee, "As Covid-19 Vaccine Microchip Conspiracy Theories Spread, Here Are Responses on Twitter," *Forbes*, May 9, 2021, https://www.forbes.com/sites/brucelee/2021/05/09/as-covid-19-vaccine-microchip-conspiracy-theories-spread-here-are-some-responses/?sh=591fde1e602d.

29 Bethany Dawson, "20% of Americans Believe the Conspiracy Theory That Microchips Are Inside the COVID-19 Vaccines, Says YouGov Study," Insider, July 18, 2021, https://www.insider.com/20-of-americans-believe-microchips-in-covid-19-vaccines-yougov-2021-7.

30 AP News, "Facebook Puts Warning on Virus Video by Retired Cardinal," January 16, 2021, https://apnews.com/article/guadalajara-

coronavirus-pandemic-latin-america-mexico-6798d15dc713a4a8009
9c7e74d0bc283.

31   Katie Schoolov, "Why It's Not Possible for the Covid Vaccines to
     Contain a Magnetic Tracking Chip That Connects to 5G," CNBC,
     October 21, 2021, https://www.cnbc.com/2021/10/01/why-the-
     covid-vaccines-dont-contain-a-magnetic-5g-tracking-chip.html.

32   Josh, "Here's How Facebook Tracks You in 2022 (and How to Stop
     Them)," All Things Secured, February 15, 2022, https://www.allth-
     ingssecured.com/tutorials/tracking/how-facebook-tracks-you.

33   Andrew Orr, "iPhones Have 100,000 Times More Process-
     ing Power Than Apollo 11 Computer," The Mac Observer,
     July 17, 2019, https://www.macobserver.com/link/
     iphones-processing-apollo-11-computer.

34   Trevor Wheelwright, "2022 Cell Phone Usage Statistics: How
     Obsessed Are We?" Reviews.org, January 24, 2022, https://www.
     reviews.org/mobile/cell-phone-addiction.

35   Seray Keskin, "19 New E-Commerce Statistics You Need to
     Know in 2022," Drip, May 24, 2022, https://sleeknote.com/
     blog/e-commerce-statistics.

36   Danielle Commisso, "Online Furniture Buying Still
     Going Strong, and Wayfair Leads," Civic Science,
     September 30, 2020, https://civicscience.com/
     online-furniture-buying-still-going-strong-and-wayfair-leads.

37   Wayfair, "About Wayfair," accessed April 21, 2022, https://www.
     aboutwayfair.com.

38   Wayfair, "Wayfair Unveils New Mobile App Features, Makes
     Shopping for Home from Anywhere Easier Than Ever Before,"
     November 13, 2019, https://investor.wayfair.com/news/news-

details/2019/Wayfair-Unveils-New-Mobile-App-Features-Makes-Shopping-for-Home-from-Anywhere-Easier-Than-Ever-Before/default.aspx.

39   Wayfair, "Wayfair Unveils New Mobile App Features, Makes Shopping for Home from Anywhere Easier Than Ever Before."

40   Wayfair, "Wayfair Unveils New Mobile App Features, Makes Shopping for Home from Anywhere Easier Than Ever Before."

41   Wayfair, "Wayfair Unveils New Mobile App Features, Makes Shopping for Home from Anywhere Easier Than Ever Before."

42   Aharonk, "A Traveler's Guide to the Champ-Élysées," July 28, 2019, https://www.aharonk.com/travelers-guide-to-champ-elysees/louis-vuitton-find-a-store-fr-louis-vuitton-maison-champs-elysees-1099_lv_18-10-15_ce_popup-now-yours_v2-di3.

43   Louis Vuitton, "Magasin Louis Vuitton Montaigne—Paris," accessed April 21, 2022, https://fr.louisvuitton.com/fra-fr/magasin/france/louis-vuitton-paris-montaigne?source=gmb%2FA02.

44   Lindsey Tramuta, "Review: Louis Vuitton Maison Champs-Élysées," Conde Nast Traveler, accessed April 21, 2022, https://www.cntraveler.com/shops/louis-vuitton-maison-champs-elysees.

45   Yelp, "Louis Vuitton," accessed April 21, 2022, https://www.yelp.com/biz/louis-vuitton-paris-7.

46   Lisa Eadicicco, "Apple Stores Make an Insane Amount of Money," *TIME*, May 18, 2016, https://time.com/4339170/apple-store-sales-comparison.

47   Jonathan Z. Zhang, "The Brand Advantage That Will Lure Shoppers Back to Stores," MIT Sloan Management Review, July 28, 2021, https://sloanreview.mit.edu/article/the-brand-advantage-that-will-lure-shoppers-back-to-stores.

48 Steve Dent, "Amazon Is Opening Its First Physical Clothing Store," TechCrunch, January 20, 2022, https://techcrunch.com/2022/01/20/amazon-is-opening-its-first-physical-clothing-store.

49 Sheera Frenkel, Mike Isaac, and Ryan Mac, "How Facebook Is Morphing into Meta," *New York Times*, February 1, 2022, https://www.nytimes.com/2022/01/31/technology/facebook-meta-change.html.

50 Sheera Frenkel, Mike Isaac, and Ryan Mac, "How Facebook Is Morphing into Meta."

51 Reuters, "Facebook Plans to Hire 10,000 in EU to Build 'Metaverse,'" October 18, 2021, https://www.reuters.com/technology-facebook-plans-hire-10000-eu-build-metaverse-2021-10-17.

52 Shamani Joshi, "What Is the Metaverse? An Explanation for People Who Don't Get it," Vice, March 15, 2022, https://www.vice.com/en/article/93bmyv/what-is-the-metaverse-internet-technology-vr.

53 Ryan William, "VR in Las Vegas: The Top 10 Experiences to Check Out," ARVR Tips, July 15, 2021, https://arvrtips.com/vr-in-las-vegas.

54 Ray-Ban, "Ray-Ban Stories Smart Glasses," accessed April 21, 2022, https://www.ray-ban.com/usa/ray-ban-stories.

55 Maghan McDowell, "Virtual Stores: Fashion's New Mode of Shopping," Vogue Business, November 24, 2021, https://www.voguebusiness.com/technology/virtual-stores-fashions-new-mode-of-shopping.

56 Nicole Silberstein, "Charlotte Tilbury Steps Further into the Metaverse with Virtual Shopping," Retail TouchPoints, November 17, 2021, https://www.retailtouchpoints.com/topics/retail-innovation/charlotte-tilbury-steps-further-into-the-metaverse-with-virtual-group-shopping.

57   Red Bull, "Red Bull the Edge," accessed April 21, 2022, https://www.redbull.com/int-en/projects/the-edge-matterhorn-vr.

58   Red Bull, "Red Bull Stratos," accessed April 21, 2022, https://www.redbull.com/int-en/projects/red-bull-stratos.

59   Gillette Venus, "Frequently Asked Questions on How to Shave with Gillette," YouTube, February 10, 2021, https://www.youtube.com/watch?v=dgLbxn2Dun4.

60   DrMatt357, "Gillette Fusion—Do These Really Work?," YouTube, August 24, 2018, https://www.youtube.com/watch?v=gY0TDUrGin4.

61   Gillette, "How to Shave—Shaving Tips for Men," YouTube, June 7, 2012, https://www.youtube.com/watch?v=wXgmMp2ZFtE.

62   Gillette UK, "How Gillette Razor Blades Are Made," YouTube, April 7, 2011, https://www.youtube.com/watch?v=ADaPr65n7hw.

63   AllyDz, "Gillette Razors—Types of Gillette Cartridges / Differences in Razors," YouTube, July 22, 2021, https://www.youtube.com/watch?v=6TVwhhdU3TU.

64   TikTok, "GoPro," accessed June 23, 2022, https://www.tiktok.com/@gopro?lang=en.

65   Minecraft, "What is Minecraft?" accessed June 23, 2022, https://www.minecraft.net/en-us/about-minecraft.

66   Stardew Valley, "Stardew Valley—About," accessed June 23, 2022, https://www.stardewvalley.net/about.

67   Joe Silk, "What Is the Long Tail and How Can It Benefit Your Business?," StartechUP, September 30, 2021, https://www.startechup.com/blog/long-tail.

68   Joe Silk, "What Is the Long Tail and How Can It Benefit Your Business?"

69   Neil Patel, "7 Brilliant Examples of Brands Driving Long-Tail Organic Traffic," NP Digital, accessed June 23, 2022, https://neilpatel.com/blog/7-brilliant-examples-of-brands-driving-long-tail-organic-traffic.

70   Neil Patel, "7 Brilliant Examples of Brands Driving Long-Tail Organic Traffic."

71   Bessma, "Is Zara a Fast Fashion Brand?" Curiously Conscious, July 16, 2020, https://www.curiouslyconscious.com/2020/07/is-zara-fast-fashion-brand.html.

72   Shobha Mukherjee, "How Zara Became the Undisputed King of Fast Fashion," The Strategy Story, November 9, 2020, https://thestrategystory.com/2020/11/09/zara-fast-fashion-case-study.

73   Frances Sola-Santiago, "Zara's Spring Collection Is Full of This Season's Biggest Trends," Refinery29, March 17, 2022, https://www.refinery29.com/en-us/2022/03/10907733/zara-spring-collection-2022.

74   Harriett Davey, "I'm Going to Zara for All My 2022 Outfits," Who What Wear, January 10, 2022, https://www.whowhatwear.com/zara-outfits-2022.

75   India Yaffe, "These 13 New Arrivals at Zara Are All Winners in Our Book," POPSUGAR, May 3, 2022, https://www.popsugar.com/fashion/best-new-arrivals-at-zara-48811709.

76   Salesforce, "Our Exclusive Customer Advocate Program," accessed June 23, 2022, https://www.salesforce.com/campaign/salescloud-first-access.

77 Salesforce, "Eight Substantial Statistics from the Trailblazer Community Impact Survey," Trailblazer Community Groups, accessed June 23, 2022, https://trailhead.salesforce.com/en/content/learn/modules/trailblazer-community-basics/meet-trailblazers-like-you.

78 Bill Murphy Jr., "Google Says It Still Uses the '20-Percent Rule,' and You Should Totally Copy It," Inc., September 13, 2019, https://www.inc.com/bill-murphy-jr/google-says-it-still-uses-20-percent-rule-you-should-totally-copy-it.html.

79 Macrotrends, "Amazon Revenue 2010–2022," accessed June 23, 2022, https://www.macrotrends.net/stocks/charts/AMZN/amazon/revenue.

80 Alex Wilhelm, "Why Amazon's History of IPO-Era Losses Means Little for Today's Unprofitable Unicorns," Crunchbase News, May 14, 2019, https://news.crunchbase.com/public/why-amazons-history-of-ipo-era-losses-means-little-for-todays-unprofitable-unicorns.